In Praise of
Things We Carry Still

"*Things We Carry Still* testifies to and raises important questions about service and citizenship in thought-provoking poems and prose. Each piece unpacks the meaning of familiar symbols—patches, pocketknives, pens, rifles, rugs—by linking them to the individual experiences of the soldiers and family members they belong to. This collection fills combat boots with soldiers' feet, unpacks uniforms from attic boxes, and places hats on the heads of soldiers' children and in doing so challenges readers to reflect on the far-reaching impact of war."

—Amber Jensen,
author of *The Smoke of You: A Memoir of Love
During & After Deployment*

"Featuring emerging and established voices, *Things We Carry Still* tells the story of contemporary military experience through the objects and rituals held dear to those who serve. Humorous and poignant by turn, this anthology offers something for everyone."

—Seema Reza, author of *A Constellation of Half-Lives*

"A loving tribute to *The Things They Carried*, this new collection from Middle West Press revisits the themes from the 1990 classic. There are many writers, and there are new stories and poems, but the song remains the same: sometimes heartfelt, sometimes cynical, and often portraying the warfighter as a well-loved cog in the wheel. *Things We Carry Still* is a powerful read."

—Christopher Lyke,
founding editor of the literary journal *Line of Advance*,
and author of *The Chicago East India Company*

Also from Middle West Press LLC

anthology

*Our Best War Stories: Prize-winning Poetry & Prose
from the Col. Darron L. Wright Memorial Awards*
Edited by Christopher Lyke

poetry collections

The Explosion Takes Both Legs: Noir Poems from the War in Iraq
by J.B. Stevens

Paying for Gas with Quarters: A Parent's Odyssey in Poems
by Aly Allen

Unwound: Poems from Enduring Wars
by Liam Corley

The Time War Takes
by Jessi M. Atherton

Hugging This Rock: Poems of Earth & Sky, Love & War
by Eric Chandler

Permanent Change of Station and *FORCES*
by Lisa Stice

Always Ready: Poems from a Life in the U.S. Coast Guard
by Benjamin B. White

HEAT + PRESSURE: Poems from War
by Ben Weakley

Things We Carry Still

Poems & Micro-Stories about Military Gear

Edited by Lisa Stice & Randy Brown
Middle West Press LLC
Johnston, Iowa

Literary Anthology / Military Uniforms / Veterans

ISBN (print): 978-1-953665-25-6
ISBN (e-book): 978-1-953665-24-9
Library of Congress Control Number: 2023941613

Middle West Press LLC
P.O. Box 1153
Johnston, Iowa 50131-9420
www.middlewestpress.com

Special thanks to Aiming Circle patrons
James Burns of Colorado Springs, Colorado
Nathan Didier of Cedar Falls, Iowa
Tim Lynch of McAllen, Texas

Your support helps publish great military-themed writing!
www.aimingcircle.com

"Gear adrift ... is a gift."
—received wisdom
from the U.S. Marine Corps

CONTENTS

Keepsakes, Snapshots & Souvenirs

Weapons, Vehicles & Equipment

Discussion & Writing Prompts

Foreword
by Vicki Hudson

In 1990, I was a senior U.S. Army first lieutenant with no actual line experience when Kuwait was invaded. My phone rang with a "Raging Bull" message, which was the code word when I answered.

Soon enough, I was at Fort Stewart in Georgia with the 351st Military Police Company, preparing for deployment. We languished several weeks waiting for our actual departure overseas date. My driver and I—Sgt. Mahon, who been a cadet until the deployment orders were issued, had temporarily reverted to NCO rank—were scavenging in a warehouse for useful items.

A short Civil Affairs major scurried over, her hands full with Army-issue pilot survival knives. "Take these," she said, her voice laced with urgency and concern. "Take these, maybe they'll be useful." We gratefully accepted.

One knife was mine; the other went to Mahon. I delivered one each to my platoon's four squad leaders.

By the end of Desert Storm, the knives' fresh white-pink leather grips were worn smooth, and burnished burnt-orange.

I'd used my survival knife every day, one way or another. It even traveled with me to other deployments, and other wars.

Long after my retirement, I picked it up from a shelf in my office. Now, when I hold the knife, it seems to come alive in my hands. So many stories, experiences, and witnessed events. The knife's edge represents the thresholds from civilian to soldier, from peace to conflict, from inactive-duty to deployment, from home comforts to austere areas of operation.

All this, from a common piece of equipment, one I carried through 33 years of military service, randomly issued by the Mobilization Log Officer one day in October 1990.

This anthology project was inspired by a realization that other veterans and military family members likely possess similar objects. Bits of uniform and equipment. A badge, insignia, or patch. Maybe a footlocker or a rucksack. So many things we are issued, wear, and carry become invested with sweat and blood and tears.

There are many stories within these pages. Some are told as prose, some as poetry. As with other war stories, each can be considered "true," even if some parts are fiction. Each is an act of courage, first in writing and then in sharing.

We carried more than boots or knives, or photos or letters, or P-38 can openers. We carried our stories. Now, we share them with you.

"Rock steady, rock on."

—*Vicki Hudson*
Lieutenant Colonel, Retired
Army of the United States

Camouflage, Clothes
& Military Decorations

Battle Dress Uniform
by Karen Skolfield

Ammunition belt. Corset of bullets.
The pattern woodland, cloud brown,

eye of the deer, beach to jungle,
a soldier must be ready, *deployment*

from the French *to unfold*. Fern curl,
frogskin, the six types of shadow,

not just mud but muck. Painted desert,
mouth of the cave, infrared dyes,

the latest science, radiation signature
of a leaf. Limbtwist, tree bark, a soldier

entrenched, low crawl, night goggles,
better than armor, a uniform

that knows the language of owls.
Blink and the woman ahead turns

to brush. Shoulder to shoulder.
Branch against another branch.

Buttons that could deflect bullets,
or so we'd been told.

Karen Skolfield's book Battle Dress *(W. W. Norton) won the 2020 Massachusetts Book Award in poetry and the Barnard Women Poets Prize. Her book* Frost in the Low Areas *(Zone 3 Press) won the 2014 PEN New*

England Award in poetry. Skolfield is a U.S. Army veteran and teaches writing to engineers at the University of Massachusetts, Amherst. Visit: www.karenskolfield.com

Uniform of the Day
by Mark Fleisher

A month after my mid-September 1967 arrival in South Vietnam, I received a "welcome present" from the U.S. Air Force: Airmen First Class would now be called "Sergeant." It was a lateral promotion—the number of stripes and the pay did not change. But I was now a non-commissioned officer and the word "sergeant" sounded far more muscular than "airman."

After getting acclimated and handling a few routine assignments as a combat news reporter for the 7th Air Force Directorate of Information at Tan Son Nhut Air Base near Saigon, I was deemed ready for the "big time." The assignment: reporting the Air Force role in a major operation involving the deployment of some 10,000 Army combat troops from a remote airstrip at Sông Bé near the Cambodian border.

Outfitted for action. I flew into Sông Bé on a C-123 "Provider." I wore combat fatigues, with pants bloused over jungle boots, helmet, and a web belt around my hips. Hanging from the belt was a canteen, and an ammunition pouch to reload, if necessary, the .38 caliber revolver snugly tucked into a leather holster. Somewhere along the line, I managed to scrounge a pair of goggles, a necessity in the swirling red dust of Sông Bé stirred up by the constant prop-wash from Air Force transports ferrying troops. And although it wasn't Air Force issue, I stuck a white plastic spoon into a shirt pocket.

Admittedly, the .38 caliber holstered on my right hip did not give me a ton of confidence. It surely wouldn't be a match for the AK-47s or Rocket-Propelled Grenades of the Viet Cong if they decided to come off the nearby hill. They serenaded us with mortar music every night.

Fortunately, I didn't find out.

Mark Fleisher has published five books of poetry and prose. He earned a journalism degree from Ohio University, and worked as a newspaper reporter and editor in upstate New York and Washington, D.C. A U.S. Air

Force veteran, he spent a year in Vietnam as a combat news reporter and received a Bronze Star for meritorious service. He is based in Albuquerque, New Mexico.

Co$t of War
by Maggs Vibo

WAR
the
price
tag
you're
it

Maggs Vibo is an author, scholar, visual-poet, and artist enchanted with folklore studies lurking in the fringes. When she isn't obsessing over poetry, the U.S. Army veteran draws inspiration from the Hawaiian island of O'ahu, where she works as an education coordinator. Image by Maggs Vibo.

Gig Line
by Benjamin B. White

It was Coast Guard Indoctrination at its best—

The explanation of the gig line:
>>> The edge of the shirt
>>> Lined up with the belt buckle
>>> Lined up with the trousers' fly.

Top-notch training
For a group with an average
Of nine years of service—

Even the prior-service Air Force guy
>>> Let his sarcasm show:

"You would have had
To have been in the military
To know that."

Benjamin B. White was pursuing a career in baseball when a worn-out shoulder diverted his life's course. After an enlistment in the U.S. Army, White ended up as an officer in the U.S. Coast Guard. He served 22 years. Now retired, the poet and novelist lives near Albuquerque, New Mexico. He is the author of more than eight books, including Always Ready: Poems from a Life in the U. S. Coast Guard *(Middle West Press, 2022).*

Boots: *Origin < Old English,*
remedy, fortunate
by Karen Skolfield

My platoon a loose group cross-legged
on the ground, boots off, bootblack tins,
the last notes of Taps clearing.

Somewhere a flag folded once and again.
In memory it never rains but smells of rain.
Wool socks, wax smudge, leather.

Eight weeks to callous up, break in boots.
"An instrument of torture used
to crush the leg and foot"

warns the dictionary's 2nd definition.
The voices of women trade shoeshine tips.
DS does mail call. We pretend not to care.

Names called or not and we tend to our boots.
Herbie's mom sends cookies to share.
Carson the boot savant, how hers gloss,

then glow, and we admire.
Lean in and it's a face dark and distorted,
some soldier I've never seen.

Karen Skolfield's book Battle Dress *(W. W. Norton) won the 2020 Massachusetts Book Award in poetry and the Barnard Women Poets Prize. Her book* Frost in the Low Areas *(Zone 3 Press) won the 2014 PEN New England Award in poetry. Skolfield is a U.S. Army veteran and teaches*

writing to engineers at the University of Massachusetts, Amherst. Visit: www.karenskolfield.com

Uniforms
by Circe Olson Woessner

Outside,
Dressed in soft greens and browns
Smoking cigarettes
Laughing and talking
The men load Conexes
Inside, I'm silent.
Sewing ...
The fabric is stiff:
beige, cream and brown
Not the woodland I'm used to
Forest or desert?
Cold War or Hot?
I examine my handiwork
Not bad, considering ...
Our PX has no green, black, or brown thread
I'm stitching with purple
Nametags on chests
Patches on arms
Who cares how you look in war?
Inside, I'm critical.
This thread is garish; my stitches are crooked ...
Outside,
The men stash their gear
Talk tactics and bullshit
It's what they've trained for—
This shiny new war
Ouch!
I suck my fingertip ruefully
... Ironic how the first bloodstain on this uniform
Is mine.

Circe Olson Woessner is the founder of the Museum of the American Military Family & Learning Center in New Mexico. As part of a U.S. Army family, she, her husband, and two sons moved 18 times in 20 years. Between 1988-1991, the family was stationed in Bad Hersfeld, Germany, which at the time was the "Outpost Farthest East in the Free World." As a freelancer, she has written columns for several New Mexico newspapers. She now generates content for 10 blogs, a podcast, and an on-line comic. She has produced multiple anthologies, including On Freedom's Frontier: Life on the Fulda Gap, Schooling with Uncle Sam, *and* Host Nation Hospitality.

Boots in the Mud
by Paul Hellweg

Shivering in cold rain
water dripping off poncho
standing in soupy mud,
you watch and contemplate
as five pairs of empty boots
are placed next to five
M16 automatic rifles
each with a helmet on top
all in a neat row, stuck in the mud.

Emotionally drained,
you want to remember and honor
but mostly you're just grateful
and pray your boots
moldy and mildewed
remain where they are,
on your feet.

Paul Hellweg is a U.S. Army veteran and freelance writer. His poetry has been nominated for the Pushcart Prize and the Best of the Net Awards. His first collection, Ode to a Drunken Muse, *was published by Alien Buddha Press. A collection of Vietnam War poems,* When Eagles Vie with Valkyries, *is forthcoming from Main Street Rag Press.*

excerpt from "your squad leader writes haiku"
by Randy Brown

Take care of your feet.
Dry socks are better than sex
out here in the field.

Randy Brown embedded with his former Iowa Army National Guard unit as a civilian journalist in Afghanistan, May-June 2011. A 20-year veteran with a previous overseas deployment, he subsequently authored the poetry collection Welcome to FOB Haiku: War Poems from Inside the Wire (2015). *He is the author or co-editor of multiple books and anthologies, including* Why We Write: Craft Essays on Writing War. *On social media, he often still writes about military themes, topics, and literature under his original mil-blogging pseudonym "Charlie Sherpa."*

D.C. al Coda
by Siren Hand

"As of July 16, 2021, 563 individuals have been arrested and charged in federal court in the aftermath of the insurrection at the US Capitol. At least 82 (14.6 percent) of those arrested are individuals with military backgrounds predominately affiliated with the Army and the Marine Corps—more than double the percentage of servicemembers in the U.S. population (approximately 6.1 percent)."
—*"The Alt-Right Movement and National Security,"*
Parameters *Vol. 51, No. 3 (2021), Army War College Press*

"Drill Sergeant, can you burn my flag?"
I wave away his smirk and motion, *Give it here.*
The Private pries the Velcro flag from his uniform,
fuzzy from wear (fuzzier than his rank),
embroidery tension-frayed from everyday tear.
The sparkstart of a lighter
a flame
        a flashpoint—
This is one thing, among us, I say,
but be careful who's watching:
we wouldn't want them to get the wrong idea.
It's one thing to burn the edges of this patch,
        make it good as new,
        acceptable for wear.
It's another to burn it out of boredom.
This is your flag, too. Care for it as you need to.
The questioning refrain:
and when people burn it for protest?
Covered under free speech. All of it:
The right to burn a flag
        to kneel with it,
        to fly it upside down in distress—

If a protestor feels there's need for it.
This cloth is voice, and presence, and power.
If not for all, then for whom?

It's 12:49 p.m.
My heart pounds
the drumroll of another Civil War
in my throat
in hours of/and seconds
in prayer
in refrain
       beating like a flagpole on the Capitol steps
       and there is no place for this type of wrong:
this spark of a flashpoint,
       a flame,
       a warning:
This is one thing, among us,
but be careful who's watching
some dead flag parades the halls as a living victor.
I wonder if the hands of Clio's clock stopped
       if she watched from the hallways of the House,
wonder if her gaze was in glee or horror,
       if it was some rebirth of a nation, again.
How many has she midwifed?

Private, are you asking because in August
A Black football player kneeled
       (kneeled: as in protest
              as in prayer
              as in reverence
              as in acknowledgement)
instead of burning it, or putting his hand over his heart,
instead of complying to violence?
Some considered this disrespect to the Flag the greatest offense—

never like recoloring it black and blue
(as in brutality, as in bruising
from the finger-deep press to find a pulse
of the Black cadaver)
Some remark this as long overdue.

This cloth is voice, and presence, and power.
If not for all, then for whom?
Give a name to the distress.
Signal however you can.
Make your grief unmistakable—
        your questions, unavoidable.
The sacrifice of symbols is   a sacred voice
People will always judge you for it
              seek ways to invalidate
it. You. Your life
              heart   beating in refrain:
This is your flag, too. Care for it as you need to.
      burn
            make it good as new,
        acceptable for wear.
You have the right.
if a protestor feels there's need for it.

I emailed the Architect of the Capitol to ask
if any clocks were broken during the riots.
"None of the historic clocks were damaged on January 6,"
A sanitized statement, without the metronome of pulse or weight
as if Time kept going,
        with or without the whole Nation behind it.

The thread from his Flag shrinks from the heat
coiling tightly, blackening into crumbling

rubbing into good-as-new
into January 7th,
       some refrain of a spotless nation
       put back, just as it was: untouched.

Siren Hand is disabled, retired U.S. Army veteran who uses writing to articulate and bridge their military experiences with civilian communities. They are currently studying creative writing and medical sociology at Indiana University Purdue University Indianapolis, with the goal of passing along to others the tools of healing that have helped them cope with their own battles. Their work has appeared in publications such as After Action Review, genesis *arts & literary magazine,* THAT *literary magazine,* Thirteen Bridges Review, *and* RockPaperPoem.

A New Set of Khakis
by Reinetta "Van" Vaneendenburg

Service dress blue was my choice to wear at our young friend Ashton's 2018 promotion to chief onboard USS *Abraham Lincoln* (CVN-72). Frank disagreed because most would be in khakis, plus the showy blue suits would be too hot for Norfolk, Virginia's sweltering September.

This would be the first time we've worn khakis, the Navy's go-to tan shirt and slacks, since we retired: Frank from Norfolk's NATO command in 2000 and me from the Reserves in 2016. While I have khakis in a range of sizes boxed in the attic, he didn't keep any.

We watch a destroyer stand out, relishing the Chesapeake Bay expanse as we linger at our kitchen table.

"Miss it?" I ask.

"No, not the Navy, but I'm looking forward to seeing Ashton's carrier," Frank says. "Ashton wants us in uniform but I can't see blowing a hundred bucks on them."

I nod. "That's why I suggested blues. How about trying Little Creek's thrift shop?" "Try," as I explain to my dubious husband, because its supervisor is a curmudgeon who won't sell uniforms to retirees. That's her edict, not the Navy's.

Frank returns toward sunset with a bag of khakis, aglow with success.

"Once she started the 'not to retirees' spiel," Frank says, "I pleaded with her. Did she want me to look shabby?"

We laugh and Frank concludes with, "Then, the old crone melted and was most solicitous."

Tuesday morning, Frank and I check each other's uniforms, an ingrained habit over the thirty-some years of military service and marriage. Mine's a little too tight, his a little too loose, but they'll do for today.

All eyes will be on the 29 new chiefs.

As they should be.

Standing tall in their new khakis.

Reinetta "Van" Vaneendenburg is a 36-year retired U.S. Navy veteran who explores through hybrid forms issues such as identity, disability, and history. She is documenting the transition of women in uniform from support to warfighter roles, drawing on her own and others' military experiences. She was the 2021 National Endowment for the Arts Military Veteran Fellow for Virginia Center for the Creative Arts Residency. Her essays have appeared in The War Horse Journal, *as well as the anthology series* Sisters in Arms: Lessons We've Learned.

Recall
by Aly Allen

 don't wear camouflage anymore,

not because it keeps civilians from seeing   me,

but because

    they already can't.

I'm imitating their perception,
Or perhaps I'm—   projecting.
The crazy mumbling veteran.   Proselytizing
    nonsense.
Drinking liquor from a bag, holding my hand out.   Proffering
    nothing.
They make bets predicting my suicide.

The projectiles still wake me
    at night, at least

The soft rumble of a rocket exploding nearby
gently sways me from sleep.
The whistle of its comrade jars me awake
before the air raid sirens scream,
we scurry to the bunker,

                the memory of us.

It's not smoke exactly,
rather   cooked dirt.
Flash-fried to the perfect temperature.
The sulfur hesitates after someone strikes a match.

21

I don't wake up in the bunker

                                        And I never sleep

inside it.

Aly Allen is a neurodivergent parent and U.S. Army Operation Enduring Freedom (OEF) veteran diagnosed with Post-Traumatic Stress Disorder (PTSD). Her kids are also neurodivergent. She served in the U.S. Army as a broadcast journalist for five years, deploying to Afghanistan once with the U.S. 10th Mountain Division. Aly recently began gender-transition, after years of sobriety and therapy. Her debut poetry collection Paying for Gas with Quarters *is available from Middle West Press.*

Dress Pants
by Benjamin B. White

The Tropical Blue Long uniform
Required blue polyester pants,
But there were never enough
In the right size available
Through the Coast Guard Exchange
                   Supply lines,
                   But the organization

Was built on a reputation
Of finding solutions
And the Air Force pants
Were the same color
                   And polyester was polyester

Although if you looked closely,
You would notice the USAF polyester
Was a different blend,
A different material,
So wearing them
Was not authorized—

                   Until necessity
                   (and unauthorized practice)
          Changed policy.

Benjamin B. White was pursuing a career in baseball when a worn-out shoulder diverted his life's course. After an enlistment in the U.S. Army, White ended up as an officer in the U.S. Coast Guard. He served 22 years. Now retired, the poet and novelist lives near Albuquerque, New Mexico. He is the author of more than eight books, including Always Ready: Poems from a Life in the U. S. Coast Guard *(Middle West Press, 2022).*

Wearing a "Bucket"
by Reinetta "Van" Vaneendenburg

This is a collection of scuttlebutt from old salts in the sea service about a peculiar uniform item: the two-tone female dress hat, circa 1942-2018. Females at boot camp in the 1970s dubbed the ungainly, rigid hats "buckets" because of their appearance when upside down. Sailors joked that such headgear was designed to catch vomitus when "the girls" became seasick.

The term was adopted at Officer Candidate School, possibly carried over by the Naval instructors, most of whom were chiefs. That's what I heard it called when I was an officer candidate in 1979.

Another nickname of that uniform article referred to a distinguishing feature of its ribbon. The length of black grosgrain circled the hat's base and overlapped in the back. The ribbon's ends were cut on the diagonal, forming an upside-down "V" that hung slightly below the back brim. I remember hearing "split tail" yelled my way by sailors and having to ask a pal what it meant. I wondered if it was, like the feminized bucket, an anachronism, something no longer authorized

The Urban Dictionary lists "split tail" by popularity. I was dismayed to see such recent usage, from 2004 to 2015. The submissions seem to be written from a male's perspective.

Since 1979, my response to such harassment was to "cage my eyes" and keep on working. We were taught to march as trainees, with our eyes focused forward, using our peripheral vision to maintain our position in the ranks. The regiment might perform a "wheel" to turn, but each head looked forward, trusting the lead to keep us safe. So too did I trust, trust that by ignoring such repetitive, offensive averment of sexual superiority, it would wear itself out, flame out, die out, be cut or kicked out.

I was wrong.

Reinetta "Van" Vaneendenburg is a 36-year retired U.S. Navy veteran who explores through hybrid forms issues such as identity, disability, and history.

She is documenting the transition of women in uniform from support to warfighter roles, drawing on her own and others' military experiences. She was the 2021 National Endowment for the Arts Military Veteran Fellow for Virginia Center for the Creative Arts Residency. Her essays have appeared in The War Horse Journal, *as well as the anthology series* Sisters in Arms: Lessons We've Learned.

Patches
by Kate Lewis

Safety is the scratch of Velcro, the squadron patches tugged from an olive drab flight suit in the depths of the night. I've mostly stopped staying up for night flights now, trusting my husband's skill and good weather and the luck that is often better, in the end, than being good. I'm used to tucking our children into bed on my own, his pillow cool beside me as I climb into my empty bed. We don't text when he takes off for the skies or when he makes it safely back to the ground, his boots once more planted upon the earth. As he returns to our hushed home, every window darkened, only one lamp left on to light his way, he tries to keep quiet, to muffle his footfalls and greet the dog with a whisper. In the hall beyond our room, he stops at the laundry and peels his flight suit patches from his arms and chest, the colorful outward record of who he is, where he's flown. No matter how silent he's been, the soft scratch always pulls me back from the brink of dreams, the sound reaching deep into my sleep to tell me just one thing. *Home.*

Kate Lewis is an essayist, poet, and military spouse who lives along the tidewaters of Coastal Virginia with her U.S. Navy aviator husband, their two young children, and a mischief-making dog. She began her career as a celebrity publicist in Los Angeles, and served for more than a decade as a PR consultant to veterans' organizations. Her essays have appeared in The New York Times, The Washington Post, Good Housekeeping, *and* Men's Health. *She was recently named a poetry finalist in the 2023 Col. Darron L. Wright Memorial Writing Awards, administered by the literary journal* Line of Advance. *She is a Perry Morgan Fellow at Old Dominion University, Norfolk, Virginia, and is writing a collection inspired by Penelope's role in* The Odyssey *and invisibility in* The Iliad. *She writes a Substack newsletter titled* The Village. *Find her on online:* @katehasthoughts

The Cover (or, The Hat)
by Andria Williams

With Susanna, now 11

When I was two-and-a-half years-
    old,
My dad suddenly brought out all
    these wonderful things!
There were gloves, and stiff vests,
    and boots I'd never seen.

But my favorite was the hat.
It flapped in the breeze!
It flipped over my eyes, and the
    dust made me sneeze.

I loved that hat!
I strutted all around.
It was like playing with a secret
    treasure I'd found.

But then everything went back into the box, even my hat.

I asked for the hat, because it was a word I could say

But then one day—
when I was almost three,

the hat was gone,

and so was he.

27

Andria Williams is the author of The Longest Night *(Random House, 2016), a novel set among U.S. Army families in Idaho in the late 1950s. It was chosen as Amazon's "Debut Novel of the Month" for January 2016. A new novel,* The Waiting World, *is forthcoming from MilSpeak Books. She holds an undergraduate degree in English from the University of California, Berkeley, and a Master of Fine Arts in Creative Writing at University of Minnesota. Williams is a founding editor of the* Military Spouse Book Review, *and a former editor-in-chief at* The Wrath-Bearing Tree. *She and her family live near Colorado Springs, Colorado, where her active-duty U.S. Navy husband is stationed.*

Photo by Andria Williams

Floppy Hat
by Barbara J. Eikmeier

He wears it on the boat, the chin strap pulled tight keeping the wind from flinging it into the lake. The floppy hat's desert camouflage was once foreign to us—brown and tan with black flecks, "like chocolate chips" the soldiers said.

In the Big Red One, he went to Saudi Arabia/Desert Shield. Right before Christmas. He wrote home: "No cowboy-style hats allowed. We wear them floppy."

He sent film. We took it for 1-hour processing, watching as the pictures shot out into a scattered pile of images: Abrams tanks. Dusty tents. Camels. Sand dunes. But none of him.

For two hours he waited in line for a 5-minute call home. I said, "I'll develop your film, but only if there is a picture of you on each roll."

More film arrived. This time, he was there. His high-and-tight haircut covered by a floppy desert hat. The strap pulled under his chin to keep the wind from blowing it away. His eyes looking into the camera. To me. To us.

We heard on the radio that the ground war had begun. Into Iraq. Into Desert Storm.

We heard when it ended. Onto Kuwait. Occupation.

We waited for him to come home.

With a child in each arm, he kissed me. He put the floppy hat on his son's head and said, "I'm home." He tightened the chin strap.

He still wears it now, but only when he's out on a boat.

Barbara J. Eikmeier is the spouse of a career soldier. She held down the homefront with two children through Operations Desert Shield/Storm, Enduring Freedom, and Iraqi Freedom. She has established homes across the country and overseas, including a tour in Korea. She is a professional quilter and published author of several books and magazine articles.

While in Uniform
by Lisa Stice

       your cover
pulled low   hides
       your eyes
       no kiss
           no hug
we say
      good-bye

Lisa Stice is the author of three full-length poetry collections, FORCES *(Middle West Press, 2021),* Permanent Change of Station *(Middle West Press, 2018) and* Uniform *(Aldrich Press, 2016), and a chapbook,* Desert *(Prolific Press, 2018). She currently serves as poetry editor for* Inklette Magazine, *as well as the Military Spouse Book Review. She writes for the Military Spouse Fine Artists Network (Milspo-FAN), and is an associate editor for Middle West Press. She currently lives in North Carolina with her U.S. Marine husband, daughter, and dog Seamus.*

Inspectable Item
by Bettina Rolyn

The list of seven Army Values was crumpled in my cargo pants side pocket. It was an inspectable item—the paper, that is, not the values—whose definitions we were induced to memorize as part of our socialization in basic training. But the spirit versus the letter versus the paper printing of said values was a story in and of itself. Who came up with these definitions, after all? Tautologies abounded. Just as the carbon corners crimped from so much rubbing against other pocketed items and friction over time, so too did my assumption of such qualities as respect and duty warp and wane with exposure to the greater system in which they were embedded.

It was a nice idea that by mandating the physical portage of these values—which were supposed to govern our lives within this institution—we would thus embody them, enact and even incarnate them. But the ink didn't ooze further than onto the inside of the camouflage pantleg, and there was no guaranteed osmosis from page to person. Eventually, I think some documents were laminated, guaranteeing their lack of effectiveness, encased as they were in little plastic sarcophagi, which added weight but not seriousness.

Bettina Rolyn was a writer and traveler starting at a young age, and lives now as a freelance German-to-English translator, creative writer, and editor in Berlin. Her essays and poems have appeared in The War Horse, The Wrath-Bearing Tree, *and* Berlin Stadtsprachen Magazin, *among others. She enlisted in the U.S. Army (2003-2007) as a Persian linguist. After quitting a career in the defense industry, she studied theology in Stuttgart, Germany and Vienna, Austria. She has a Master of Fine Arts in Creative Nonfiction from Carlow University, Pittsburgh and is finishing a memoir.*

Flight Suit
by Layle Keane Chambers

I visited in May for no reason, just needed a day at Lake Amistad.
We pitched a tent on nothing but rocks. They felt warm
and therapeutic. We dripped easily
between the crevices, sharing watermelon
and a bottle of Pool Boy
Rosé in a plastic bottle, brilliant.
The dogs ran too far, but there was no one
and we reveled in our possession of the shore. We
found an old foundation wall with bricks
stamped St. Louis

I visited in August
just because you were still in Texas and I could.
You invited me to see the flight line, maybe watch you take off
from the balcony outside Silver Wings where I wanted
to buy chips and a soda like a teenager, but didn't

I sat at an empty white Formica table
and watched young men and women in their flight suits,
remembered walking into Drunken Fish with you in yours,
how I said it attracted more attention than high heels and a mini skirt.
We laughed at the idea of you in high heels and a mini skirt
and the waiter showed us to a table where people wouldn't stare,
and then he paid our check.

I wandered outside thinking it might be time,
leaning on the warm metal railing and looking out over gathered aircraft,
lines of T-38 Talons and the T-1 Jayhawks that you describe
as flying a concrete box.
I remembered crying, relieved, the day you told me
you were going to vector heavies.

Lost in my trance of what a different choice may have meant, I hear *'hey mom!'* from the tarmac and you are waving.

Layle Keane Chambers is a teaching artist currently living in Folly Beach, South Carolina. She received her undergraduate fine arts degree from Ithaca College and master's degree from University of Texas at El Paso in Theatre Arts. She is a professor emeritus of theatre at Doña Community College, Las Cruces, New Mexico; a Master of Fine Arts Candidate in Creative Writing at New Mexico State University; and a proud U.S. Air Force mom.

Maternity BDUs
by Mary Senter

I was working vertical construction in the Air Force when I decided to become a single-mother-by-choice. It was an unpopular decision to everyone but me. To me, it seemed perfectly reasonable. I was in fantastic shape, I had a steady income, maternity leave, housing, medical insurance, and a sense of security. I came from a broken home and a subsequent blended family and marriage was never something that I saw for myself. But I wanted to be a mother.

I loved being pregnant and I loved wearing maternity BDUs—the forest-camouflage "Battle Dress Uniform." I slipped on the stretchy-waisted pants, laced up my 16-inch lineman's boots, bloused my pants over them, and buttoned up the smock-cut shirt. The only downside was that, because there was no belt, I had nothing on which to hook my Leatherman, which was a staple of my uniform and a constantly used tool. So was my tape measure and ear protection. I missed the Leatherman the most, though, and felt naked without it.

Mary Senter creates in a cabin in the woods on the shores of the Salish Sea, in the U.S. Pacific Northwest. The U.S. Air Force veteran has earned certificates in literary fiction-writing from the University of Washington, and a graduate degree in strategic communication from Washington State University, Pullman. Her work can be found in North American Review, El Portal, Drunk Monkeys, Ponder Review, Cleaver, *and elsewhere. She is the founder of Milltown Press. Visit: www.marysenter.com*

Non-Essential Equipment
by Jehanne Dubrow

The dog and I are first among those things
that will not be deployed with him. Forget
civilian clothes as well. He shouldn't bring
too many photographs, which might get wet,
the faces blurred. He only needs a set
of uniforms. Even his wedding ring
gives pause (what if it fell?—he'd be upset
to dent or scratch away the gold engraving).
The seabag must be light enough to sling
across his shoulder, weigh almost nothing,
each canvas pocket emptied of regret.
The trick is packing less. No wife, no pet,
no perfumed letters dabbed with *I-love-yous*,
or anything he can't afford to lose.

Jehanne Dubrow is the author of nine poetry collections, including most recently Wild Kingdom *(Louisiana State University Press, 2021), and two books of creative nonfiction,* throughsmoke: an essay in notes *(New Rivers Press, 2019) and* Taste: A Book of Small Bites *(Columbia University Press, 2022). Her third book of nonfiction,* Exhibitions: Essays On Art & Atrocity, *was published by University of New Mexico Press in 2023. Her writing has appeared in* POETRY, New England Review, Colorado Review, *and* The Southern Review. *She is a professor of creative writing at the University of North Texas.*

Party Dress
by Lee Anne Gallaway-Mitchell

25th Fighter Squadron, Assam Draggins, Osan Air Base, Korea—She wears a Kelly Green party dress, a dragon embroidered down the side. It matches her husband's party suit, the same dragon roaring down his legs. So fresh and so pretty, like a sorority girl, she tilts her head and plays carelessly with the Kim Jong Un bobblehead sitting on the bar at the fighter squadron.

"Now, who's this little fella?"

The young woman's husband tells her that the bobblehead is the leader of North Korea. She shrugs and takes a sip of her beer. The Korean Demilitarized Zone (DMZ), a 250-kilometer-long active minefield, sits 56 kilometers away.

Just around the corner, next to the Taco Bell, the Base Exchange sells small commemorative DMZ plaques with "original" barbed wire.

You get fitted for the party dress within weeks of arriving in Korea. The party dress resembles a retro-style flight attendant uniform: Short, form-fitted. You'll have a name patch made to match your husband's, and a Texas flag to go on your shoulder. A tiger will breathe fire down your thighs, too.

Your husband's party suit is a bell-bottomed throw-back to 1960s Thailand, the social dress for fighter pilots once stationed there. On his back, a cartoonish A-10 Warthog flies over a voluptuous mountain range highly suggestive of a naked woman's body. At the Officers' Club, he has to wear a leather biker vest over it. Because of the dress code.

Your suit will be ready in one week, they tell you. You keep watching the news. You see the failed rocket launch attempts from the north. You pick up the party dress at the tailors outside the gate in Songtan. It fits you perfectly.

Lee Anne Gallaway-Mitchell lives and writes in Tucson, where she received a Master of Fine Arts in creative writing from the University of Arizona. Her essays have won The Florida Review Editor's Award, the Arts & Letters Susan Atefat Prize for Creative Nonfiction, and, most recently, the Boulevard Magazine Nonfiction Contest for Emerging Writers. She is at work on an essay collection, Campfollowers.

Army Ball
by Abby E. Murray

You've outgrown the prom,
the men I mean, not us, the wives,
who spend hours buffing time
from our skin and dazzle
in pearls and tennis bracelets
clipped like medals to our bodies:
the OIF amethyst, OEF diamond studs,
SFAT cashmere, reintegration pearls.
Some new wives miss the mark,
overshoot the dress code
and show up in wedding gowns.
They pick at the crystals, the ruching.
At our table, your jaw is softened by gin
and lost time, that year before Iraq
when Black Hawks dropped you
into the unarmed mountains of Alaska,
a simulated war with its certain end.
The colonel's wife talks to me
about her family law practice,
eight years untouched now
on account of her boys and the traveling.
I want to hug her but we've just met
and I know she is being kind.
I'm wearing polyester, faux-leather mules,
pinned my hair up in the car.
We are saved by shushing
for the grog ritual: men of different ranks
come forward with liquor bottles so large
they represent entire wars:
dark rum for the jungles of Vietnam,
canned beer for Afghanistan.

A bowl the size of a bus tire is filled
with two hundred years of symbolic booze
and we hoot and clap when the men
take long drags from each bottle or can,
we scream as if to say the weapons burn
in our throats the way they do in theirs.
Waiters come round with pitchers
and serve grog with silver ladles
polished last night, too early,
tarnish blooming in their grooves.

Abby E. Murray was the 2019-2021 poet laureate for the City of Tacoma, Washington. Her first full collection of poetry, Hail and Farewell, *won the 2019 Perugia Press Poetry Prize and was a finalist for the Washington State Book Award. Previous chapbooks include* How to Be Married after Iraq *(Finishing Line Press, 2018),* Quick Draw: Poems from a Soldier's Wife *(Finishing Line Press, 2012, and* Me and Coyote *(Lost Horse Books, 2010). She is the founding editor of* Collateral, *a literary journal publishing work concerned with the impact of violent conflict and military service beyond the combat zone. She teaches rhetoric in writing military strategy for U.S. Army War College fellows at the University of Washington, as well as poetry workshops at community centers, coffee shops, military posts, detention centers, shelters, and schools.*

Navy Pride
by Charles McCaffrey

An ensign knocks at my stateroom door, and asks if I have an extra Sea Service Ribbon that he can use for the personnel inspection starting in 15 minutes with our Destroyer Squadron Commander.

"Sure," I say, and hand him a ribbon out of my bag of military decorations.

He's back a few minutes later, nervously asking me to make sure his single row of ribbons is in the correct order. I stare at them with a look of mild confusion.

"What?" he asks in a panic.

"I can't remember if that's the Sea Service Ribbon or the Gay Pride one." There is no Gay Pride Ribbon.

"Well, which is it?" he pleads with genuine terror in his eyes.

I shrug and give him a clap on the shoulder.

"I'm sure you'll be fine either way," I assure him as I stroll out of officer's country to join the ship's crew on the fantail.

Charles McCaffrey is a U.S. Navy veteran, and an avid storyteller and writer. As a kid, he always had paper and a box of crayons with him; drawing and writing about people, places, and things both real and imagined. He still carries a pen and notebook with him everywhere he goes.

Uniform Gedunk
by Benjamin B. White

On one hand,
I understood and empathized
With the pride
Of personal accomplishments,

        But on the other hand,

The Coast Guard
Had to draw the line
Only allowing
Certain ribbons and medals
From other services

        To be worn on their uniforms

So in the end,
I had to fundamentally agree
No one in the Coast Guard

        Should wear jump wings
      Or a Combat Infantryman Badge

Benjamin B. White was pursuing a career in baseball when a worn-out shoulder diverted his life's course. After an enlistment in the U.S. Army, White ended up as an officer in the U.S. Coast Guard. He served 22 years. Now retired, the poet and novelist lives near Albuquerque, New Mexico. He is the author of more than eight books, including Always Ready: Poems from a Life in the U. S. Coast Guard *(Middle West Press, 2022).*

Bottle Opener, or "Sestina for My CIB"
by Colin D. Halloran

Shades drawn, I sit on the floor and pry
my honor's edge under corrugated rim: first
of six. Release cutting illusion of silence
drone of the muted box, unseen
where I contemplate and combat
the spin-out since coming home.

I wanted nothing more than home,
those I'd missed, but others pried
into events of seven months in combat.
I was bombarded with inquiries first,
before I could even process what I'd seen.
I greet them with a distant stare and silence.

In Uruzgan, in rare moments of silence
and solitude I would cling desperately to home:
New England's shell strewn shores, not children's blood seen
spilled on sand, a violent new day-by-day to pry
me from my dreams. Mission first.
Constant alert, plan, execute: the rhythm of combat.

MOS 11B: Infantryman. Front line combat,
my weapon's frequent bursts will lead to silence
ringing. I am not the first
to earn this badge, to risk home
to protect abstractions. It's not about fighting those who try to pry
Freedom or Democracy away from a place they've never seen;

it's protecting those few who share what you have seen,
brothers in arms, strangers made inseparable through combat
until death, or injury, or pride, or peace begin to pry

you apart. I am a recluse, an uninvited vow of silence
and isolation, an early, guilt-ridden return home.
The injury I tried to push through. Med-Evac. Not the first

injured to leave my platoon, and even though at first
I fought to stay, when the knee began to give, when doctors had seen
the damage, the incapacitation, the risk I posed, I was bound for home.
Forced to leave the violent province, newfound brothers, life of combat,
return to this life that lacks adrenal kicks, my head hung in the silence
Of guilt, pain, personal defeat, and the slow slipping away of pride.

Now I assault this six-pack, try to pry myself away from all I've seen,
from the silence of being caught between those I left in combat,
and the unknowing faces of home that once were first to know me.

*Colin D. Halloran is a U.S. Army veteran of Afghanistan. He authored
the award-winning memoir-in-verse* Shortly Thereafter, *as well as*
Icarian Flux *and* American Etiquette. *He lives and teaches in Alabama.*

Wings of Gold
by Jason Gaidis

Memories impact my mind like missiles riding a beam
It started as a childhood dream
The same dram every child has became
a goal that became reality
Something real I fought for, inside and out of me
Everything to get underway and stand alert thirty
SAR, Hellfire, and blue-water ops became the best part of me
Preparing to save life and end life became my identity
But now that I'm out, that part of me causes controversy
My aggressiveness, decisiveness, detachment and drinking
    are called PTSD
Because for me, retirement changed it all
After 20 years Uncle Sam gave me a pension, pills and a folded flag
    for answering duty's call
and my wings of gold moved from my uniform to the wall
It's all a way to rationalize the way I am
The kind of person people appreciate but few understand
as they stand and shake my hand and thank me for my service
It's fun in the movies but my reality makes them nervous
They say: get sober, get serious, bet 'better,' get a 9-to-5 and punch it
Welcome to the 'real' world, you'd better adapt to it
Time for new commanders intent and criteria
    for mission accomplishment
as I raise my daughter who will never know me as a Navy Pilot
She will learn by my wife's tears while speaking of deployment
And my enjoyment in the sea stories long told
Of the time I served in the United States Navy and wore wings of gold.

Jason "O.M.G." Gaidis earned his Master of Fine Arts from Ashland University. His writing has appeared in Adrift Anthology *from Third Street Writers;* Proud to Be: Writing by American Warriors, Vol. 9 *from Southeast Missouri State University Press;* Black Fork Review *from Ashland University, and others. Jason is a retired U.S. Navy officer and aviator. He lives and writes with his wife, Amy, in Grosse Pointe Woods, Michigan.*

EGA
by Mason Rodrigue

The Crucible was only as bad as you believe it is.
If you're going infantry
it's only just a taste.
Three days came and went quicker than I thought they
would,
almost underwhelming.
We were so physically conditioned that all the running
and jumping and fighting
was just par for the course.
They can only make you low crawl for so long,
and the cold, wet mud washed off easy enough.

Nights sucked, but I was only just learning how shitty
it is
to have to get out of warm sleeping bags
on freezing mornings
to a sun that won't help.

The pain of it is cumulative,
and the nine-mile hike in the blizzard was grueling, but
Once we began that last stretch
and marked under the
"We Make Marines" arch
I knew I'd made it.

Hiking without warming layers left me we with sweat—
shivering in the cold, and aching
both physically and emotionally

Standing at parade rest, waiting
The Drill Instructor appeared in front of me and asked

"What does this mean to you?"
I stared through him
with all the Discipline he'd drilled into me
and replied
"Everything."

He put the black Eagle, Globe, and Anchor emblem in my
hand
And shook it
Like a man
And I cried
Like a baby

Mason Rodrigue (pronounced "road-rig") was born in southern Louisiana. The terrorist attacks of September 11, 2001 led him to enlist in the U.S. Marine Corps and eventually his service in Syria as a machine-gunner. Rodrigue is the author Rock Eater *(Dead Reckoning Collective, 2023), a collection of poetry and philosophy about "what happens when a man gets turned into a weapon and aimed at nothing."*

Changing to the Major's Dress Blues
by Lisa Stice

This uniform is different from the last
ten years. This one settles the heart.
Your thick brocades of gold,
age-forgiving cummerbund,
(more bravado than anything else)
are like nearing something:
we call it, career: spent in training for
combat zones and long silences.

This is the turning point—
for all these memories of pinning ribbons,
lining up medals, turning buttons,
making certain they're straight,

so that one day we will retire them to a bag
in a closet next to my evening gowns.

Lisa Stice is the author of three full-length poetry collections, FORCES
(Middle West Press, 2021), Permanent Change of Station *(Middle West
Press, 2018) and* Uniform *(Aldrich Press, 2016), and a chapbook,* Desert
(Prolific Press, 2018). She currently serves as poetry editor for Inklette
Magazine, *as well as the Military Spouse Book Review. She writes for the
Military Spouse Fine Artists Network (Milspo-FAN), and is an associate
editor for Middle West Press. She currently lives in North Carolina with
her U.S. Marine husband, daughter, and dog Seamus.*

Boots
by Amalie Flynn

When my husband comes home, from war, *for good*, there are still checkpoints. There are those points we have to pass through each day.

Like when he sees his combat boots in a closet, lined up and empty, reminding him of all of the soldiers who are dead, now, and how it could have been him.

Like when he sees my back, again, facing him, in bed, this massive trench forming in between us, a massive trench that neither of us knows how to cross.

Amalie Flynn is a poet and the author of September Eleventh: an epic poem, in fragments *(Middle West Press, 2021);* Wife and War: The Memoir *(2013); and a collection of poetry blogs:* Pattern of Consumption, September Eleventh, Wife and War, The Sustainability of Us, *and* Border of Heartbreak. *Her writing has appeared in* American Book Review, Beyond their Limits of Longing *(2022), the* New York Times, Time, *and* Huffington Post *and has received mentions from the* New York Times *and* CNN. *Flynn has an undergraduate degree in English and Studio Arts, a Master of Fine Arts in Creative Writing, and a doctorate in Humanities. She serves as poetry editor for* The Wrath-Bearing Tree *literary journal. Flynn lives in Rhode Island with her husband and their two children. She is currently working on a poetry collection focused on sexual and environmental violence.*

Boots on the Ground
by Aaron Graham

The boots on the ground aren't boots
they're packs of young men roaming
streets eternal out-of-towners in combat
patrols come to see the valley of death
that is the cradle of civilization.
Sightseeing tours of duty:
street corner claymore mines
DIY daisy chains
IEDs built into the fucking roads.
Fresh laid blacktop
and ancient alley are equal parts
alligator snapping turtle
sandstone gape-mouth
and shrapnel swallows.
Feeling expands in the air
    like flame-plumes—
scorching in gouts.
Out past the perimeter,
terror's on every side—
front-sight tips,
rear-sight aperture—
stoned and immaculate.

*Aaron Graham is a veteran of the wars in Afghanistan and Iraq. He served
with The Marine Corps' Human Intelligence and Counterterrorism Task
Force Middle East as an analyst and linguist. His debut poetry collection,*
Blood Stripes, *was published in 2019 by Sundress Publications. His 2018
chapbook* Arabic with a Redneck Accent *won the Moonstone Center for
the Arts Poetry Prize. He lives with his family in Greensboro, North
Carolina.*

Ditty Bag
by Yvette Viets Flaten

My father's Blues are still packed away
in his B-4 Bag, the military's bulked-up
version of an Executive Suiter, as if he's
on stand-by for another TDY.

In that blue bag he brought back
the booty of his wars: Libyan prayer rugs,
hand-carved camels. Baltic amber.
Thai princess rings and a blaze of silks.

All we would hear is the thump in the hall
as he threw it in, followed by our shrieks and kisses.

Each zipper peeled back a layer of story:
of desert, of cold, of jungle. Foods, and people,
and sorties, in and out.

His Ditty Bag was smaller.
An overnighter, I guess you'd call it.
Socks, and shorts, and shaving kit; that sort of thing.
He kept it loaded, ready to go.
A Traveler's Check, folded small
like origami. A 3x5 card with some
stateside addresses. A clean, white undertop.
V-necked. Rolled up and rubber-banded.

I just found his last ditty bag,
the one he was given at the VA.
It's yellow, and has a drawstring.
Sewn by a volunteer, herself a widow.
It held his Tums, some change, and another

3x5 card with his home phone number
written in thick marker.

He held it on his lap when they rolled him
from one room to another. Gripped it with
hands that had repaired flak damage
and signed off on turbo-fans.

Gripped it with the same hands
that tossed that B-4 Bag into
the hallway of my childhood,
hands that circled my mother's
face for a kiss, hands that held
us so very, very tight.

*Yvette Viets Flaten grew up in a U.S. Air Force family, living overseas in
England, France, and Spain, and stateside in Nevada, North Dakota, and
Washington state. As a result she learned to appreciate language, travel,
history, new cultures, and new foods. Flaten writes both fiction and poetry,
and makes it a point to write every day.*

Keepsakes, Snapshots
& Souvenirs

Footlocker
by Colin D. Halloran

Each soldier is issued a footlocker—
approx. 32½" x 15¾" x 13¾"—
for personal effects:
clothes, magazines, movies, mementos
of a home they're leaving behind,
or essentials like cigarettes and jerky.
Pornography and booze have no place
in these silver-latched intimate havens,
they violate General Order 1.
Each person's is different, though
there are some items found across the board,
namely baby wipes and sunscreen,
but they say you can learn a lot about a man
from his box's contents.
My box is my own.
A 5-pound jar of Skippy,
the latest issues of
Surfing
and
Whitewater—
so at least my dreams are not left dry in the desert—
most of the remaining space is dedicated to an infantryman's anomaly:
    books.
I can't seem to go to war without Shakespeare, Uris, Whitman, Keats.
I am not a typical grunt
is what my footlocker says.

*Colin D. Halloran is a U.S. Army veteran of Afghanistan. He authored
the award-winning memoir-in-verse* Shortly Thereafter, *as well as*
Icarian Flux *and* American Etiquette. *He lives and teaches in Alabama.*

So I Was a Coffin
by Gerardo "Tony" Mena

for Corporal Kyle Powell, died in my arms, 04 November 2006

They said you are a spear. So I was a spear.

I walked around Iraq upright and tall, but the wind blew
    and I began to lean.
I leaned into a man, who leaned into a child, who leaned into a city.
    I walked
back to them and neatly presented a city of bodies packaged in rows.
They said no. You are a bad spear.

They said you are a flag. So I was a flag.

I climbed to the highest building, in the city that had no bodies,
    and I smiled
and waved as hard as I could. I waved too hard and I caught fire
    and I burned
down the city, but it had no bodies. They said no. You are a bad flag.

They said you are a bandage. So I was a bandage.

I jumped on Kyle's chest and wrapped my lace arms together
    around his torso and
pressed my head to his ribcage and listened to his heartbeat.
    Then I was full, so
I let go and wrung myself out.

And I jumped on Kyle's chest and wrapped my lace arms together
around his torso
and pressed my head to his ribcage and listened to his heartbeat.
    Then I was full, so

I let go and wrung myself out.

And I jumped on Kyle's chest and wrapped my lace arms together
      around his torso
and pressed my head to his ribcage but there was no heartbeat. They
      said no. You
are a bad bandage.

They said you are a coffin. So I was.

I found a man. They said he died bravely, or he will. I encompassed him
in my finished wood, and I shut my lid around us. As they lowered us
into the ground he made no sound because he had no eyes
and could not cry. As I buried us in dirt we held our breaths together
and they said, yes. You are a good coffin.

*Gerardo "Tony" Mena is an Operation Iraqi Freedom veteran who spent
six years in the Special Operations community with the Reconnaissance
Marines as a Special Amphibious Reconnaissance Corpsman (SARC).
During his tour in Iraq, he was awarded a Navy Achievement Medal with
a "V" device for multiple acts of bravery. His debut poetry collection,* The
Shape of Our Faces No Longer Matters, *was published by Southeast
Missouri State University Press in 2014.*

Black Footlocker
by Aramis Calderon

The black footlocker
flew, sailed, and trucked
to a land filled with
prosperity, opportunities, and dreams.

It contained another country.
One filled with
poverty, threats, and nightmares.

Sirens.
Explosions.
A call to prayer.

All this and more were
safe inside.

Aramis Calderon is a U.S. Marine Corps veteran and the author of the novel Dismount *(A15 Publishing, 2019), and the forthcoming memoir* Fugitive Son *(Potomac Books, 2024). He holds a Master of Fine Arts in Creative Writing from the University of Tampa. His current area of operations is Tampa Bay, Florida, where every week he meets with fellow veteran writers in the DD-214 Writers' Workshop.*

"who dares, wins"
by Suzanne S. Rancourt

i am sitting in the hammam with the old men
our tats
a wordless mercy—no talk—in the steam
thick as croup, branding iron sssst

our tats hang on flesh
101st, 82nd, 10th mtn, red 1, EGAs,
daggers, lightning bolts, with wings, no wings,
skulls, V-42s, k-bars—

herme's caduceus entwines my spine
we turtle down in to our shoulders and pray
for scorching steam that wipes away
burdens

mercy for my brethren,
these old men who died young
we sit on tiled benches in low light
our flesh sweats, we cough up shit. I pray

gratitude for this moment
where faded ink
tapped in code
still sings our stories.

Suzanne S. Rancourt is a multi-modal expressive arts therapist with graduate degrees and certifications in psychology, creative writing, drug and alcohol recovery. She is a veteran of the U.S. Army and U.S. Marine Corps, and is of Abenaki/Huron descent. Her debut collection of poetry, Billboard in the Clouds, *received the Native Writers' Circle of the Americas First Book Award. Her fourth collection,* Songs of Archilochus, *was recently published by Unsolicited Press.*

Flags and Sunglasses
by Charles McCaffrey

The photo sits on my desk. In it, he's posing with his team before heading out on patrol. He's in full kit standing in front of his Humvee—no smile; his eyes are hidden behind his favorite pair of Oakley sunglasses.

At the funeral, I was allowed to sit in the front row because I was friends with the family.

Military funerals are a funny affair. Not funny "ha-ha" but funny surreal. Attend a funeral just about anywhere else in the world and there will be wailing and celebrations, fist fights and drinking, gunfire and fireworks, embraces and protests. But here in America, we grieve in silence; and people make casseroles. I hate casseroles.

And at a military funeral, your brothers- and sisters-in-arms stand in silence; stone-faced, back straight and chin up. Defiant in the face of death and grief and tragedy. We don't cry in public. I didn't cry. I disassociated myself from my surroundings because I knew if I thought about this, our life and his death, for even a moment, I would break down bawling, again.

I passed through the five stages of grief every couple of minutes— denial, anger, bargaining, depression and acceptance. No, not acceptance, not yet. This had gone on for weeks. I was exhausted and numb.

His mother received the folded flag from his casket.

After the funeral, a fellow Marine who knew we were more than just friends, quietly handed me his favorite pair of Oakley sunglasses—now scratched and cracked. They sit next to the photo of him on my desk.

Charles McCaffrey is a U.S. Navy veteran, and an avid storyteller and writer. As a kid, he always had paper and a box of crayons with him; drawing and writing about people, places, and things both real and imagined. He still carries a pen and notebook with him everywhere he goes.

Silver Dollar
by Amalie Flynn

This is the moment we have to say *goodbye.*

Goodbye, goodbye for fifteen months.

And our son is only two years old, sitting in a car seat, in the backseat, in the car. And my husband is leaning over him, leaning in the car door, trying to give him a special coin, a silver dollar. And our son will not take it, putting his little hands, behind his back, *no* and *no.*

It is as if he knows, as if he knows what it means.

How I will take it, the silver dollar, take it home, and put it in a box, on a high shelf, in my closet.

Because if my husband dies, his father, if he dies in this war, it will be the last thing he ever gave him.

Amalie Flynn is a poet and the author of September Eleventh: an epic poem, in fragments *(Middle West Press, 2021);* Wife and War: The Memoir (2013); *and a collection of poetry blogs: Pattern of Consumption, September Eleventh, Wife and War, The Sustainability of Us, and Border of Heartbreak. Her writing has appeared in* American Book Review, Beyond their Limits of Longing *(2022), the* New York Times, Time, *and* Huffington Post *and has received mentions from the* New York Times *and* CNN. *Flynn has an undergraduate degree in English and Studio Arts, a Master of Fine Arts in Creative Writing, and a doctorate in Humanities. She serves as poetry editor for* The Wrath-Bearing Tree *literary journal. Flynn lives in Rhode Island with her husband and their two children. She is currently working on a poetry collection focused on sexual and environmental violence.*

Meeting in Amman
by Farzana Marie

The list boasted benefits
for 100,000 names–sons of a country
whose leader proved as faithless
as the duped-power who trusted,
handed him the list,
vanished.

Two men met in Amman
even as the list bled, as it
led the latest so-called caliphate
to doorsteps of soldiers
now called apostates.

The American colonel, retired,
tired of excuses,
could not open his mouth, once full
of promises.

Instead, the Sahwa leader opened
his pockets, vomiting dozens
of U.S. military unit coins
over the table and ground,
his voice an earthquake:

what good are these now?

Farzana Marie is a poet, writer, and stroke survivor. She lives in Albuquerque, New Mexico and has lived in Arizona, Afghanistan, Kazakhstan, Chile, and California. She served two consecutive years deployed in Afghanistan, where she also served as a civilian volunteer at a

Kabul orphanage in 2003-2004. Farzana has published her translations of Afghan women's poetry, a poetry chapbook, and a non-fiction book. She has a Ph.D. on Persian literature with a minor in creative writing from the University of Arizona, she holds a Bachelor of Science in Humanities from the U.S. Air Force Academy and Master of Arts in English from the University of Massachusetts, Boston. In 2015, she had a massive stroke in Afghanistan. Today she fights a new adversary: aphasia; loss of language skills—not intellect. She works every day to recover her ability to speak, read, and write.

Ouroboros in a Hula Hoop
by Kate Carey

"I think Carey may have been with Bravo Company, but I'm not sure. According to Dexter's diary, we had enemy contact that day and thought for a while that we had a man missing [MIA]. Eventually a body was found near where our contact originated. The company had pulled back and artillery was called in. At some point during that time, a man was discovered missing. I helped carry him out and I remember that he had a huge gash in his chest. We speculated that he was hit by artillery, but who knows? I remember it was a new man ... and Carey arrived in country February 15, 1968."

—Larry Mitchell, Bravo Co. 1967-68 (10/8/2000): www.manchu.org

My brother had about twenty bucks in his wallet when he was killed, saved for a cold beer in some dingy bar near Saigon. Mom split it between me, age nine, and my brother, eleven. He says he doesn't remember. I bought a hula hoop.

I'd not yet experienced death, and visitations, and funerals, and burials with Taps, folded flags, and 21-gun salutes. At nine, death money was tangible in ways a closed casket wasn't.

I idolized my big brother. Star football player in high school. Handsome, smart, funny. Dave let me drive his car. Okay, I sat on his lap and steered but to me, that was driving. He gave me my first rock album. The Monkees. I loved Mickey Dole.

A hula hoop.

Inconceivable now, remembering that year of death and discrimination buoyed by hope. Tet, Martin, Bobby, Olympics protests, Civil Rights Act, Men to the moon.

But metaphorical.

A circle. Life, death, life again.

Balance.

Ouroboros. A snake eating its own tail. Infinity. Wholeness.

A hula hoop that I spun around and around allowing myself infinite joy that dark, cold spring.

Kate Carey grew up in Ohio, on 88 acres and a farmhouse her mother's family had owned for generations. In March 1968, two soldiers knocked on the front door with news that forever changed their lives. Her brother Dave's life insurance helped fund her college education. Now transplanted to North Carolina, she regularly writes for Topsail *magazine. Her writing has also appeared in* Noctua, Indiana Voice, The Tishman Review, Panoply, Camel City Digest, Savannah Writers Anthology *and* County Line Journal.

I am not Superstitious
by Eric Chandler

In the 4th grade, I bought a jackknife for $4.75. I carried it everywhere. I put it in a pocket of my flight suit once I started flying for Uncle Sam. I always flew with it. After I left the service, I took it with me to the airlines.

Knives weren't welcome at the airport after 9/11. I put my pocketknife in a drawer and carried a New Hampshire state quarter instead. That coin depicts the iconic stone face of The Old Man of the Mountain. This landmark is just a few miles from where I was born. I always flew with that quarter.

After 9/11, my airline laid me off. I joined the Guard. I put the flight suit back on and carried that quarter.

In 2003, the great stone face tumbled down the mountain. I was sad that The Old Man was gone. But I was more concerned about my quarter. Was the collapse of the stone face an omen?

The New Hampshire state motto is "Live Free or Die." The Old Man falling off that cliff resembled the "Die" part. I choose to think he retired. Maybe he wanted to "Live Free" after sitting on that ledge since the Ice Age. Travel some. So, I took The Old Man coin to Iraq and Afghanistan. The Old Man was a good wingman.

Then I got old and retired, too. We went back to the airlines.

I just checked that pocket in my flight bag to make sure my quarter was still there.

The Old Man is flying with me to work right now.

I don't need The Old Man.

I could lose The Old Man tomorrow.

It wouldn't bother me.

I am not superstitious.

Eric Chandler is the author of the poetry collections Kekekabic *(Finishing Line Press, 2022) and* Hugging This Rock: Poems of Earth & Sky, Love & War *(Middle West Press, 2017). He is a three-time poetry winner of the Col. Darron L. Wright Memorial Award, and a retired veteran of the U.S. Air Force and the Minnesota Air National Guard. Over Iraq and Afghanistan, he flew 145 combat missions in the F-16. He is happiest when on a Duluth, Minnesota trail with his wife, two children, and dog Leo.*

Pink Elephant
by Jessi M. Atherton

The heat at the Baghdad bazaar
was almost unbearable
I wandered with my squad
down tables filled with things
to buy and send home or
keep as memories

I had not really done the latter, because
I thought I couldn't
care less about that place.
As I continued to find nothing,
I stopped at a table filled
with carved animals—
some made of quartz, others
of marble or wood.

You were attending the table.
You were the artist behind
all those giraffes, turtles,
hearts, and horses.
You were very talkative, polite.
When I asked your name
You said Mohammed, but also
to call you "Mo."

I asked if you ever carved elephants—
my favorite animal—
You smiled and said "Next time you come,
there will be an elephant for you."
I nodded, but did not believe.

THINGS WE CARRY STILL

How would you remember?

Two weeks passed by before
we were outside the wire again.

I made my way to your table—
You welcomed me with a smile.
I saw several elephants, newly carved
and placed in a row.
I picked one up.

"No," you said, "Not that one—
for Jess, I made special one."
You reached into a box
and revealed a carved pink
elephant, made from quartz.
I was surprised by your memory,
and touched by your kindness.

It was the perfect item
to bring home.

Pink elephant stayed with me
for the rest of my deployment in Iraq.
tucked away in a pillowcase
wrapped up in newspaper
inside my green duffel bag.

She made the long journey home
First, as we medically evacuated out of Iraq
on a flight to Germany
where I recovered for weeks
then onward back to America—
Fort Bliss! Then Fort Riley!

Six months later,
I was med-boarded out of the Army.

Pink elephant was with me
as I transitioned into being
a civilian again—I began the next chapter
of life as I re-entered college,
finishing a degree, then starting another
in Nursing.

She sat on a bookshelf full of medical texts
and self-help books,
watching me re-discover
who I was ...

Pink elephant made it into
the "things we can't live without" pile
as a Colorado wildfire threatened
our home and belongings
and we prepared for evacuation ...

Pink elephant sat on my dresser
as I nursed my three babies
all hours, day and night ...

Pink elephant followed me through
mothering milestones, career changes,
a divorce,
the rebuilding of my life ...

Many times, I have picked her up
and run my hands over
her smooth surfaces, looking at the intricate

formations of pink and white rock—
comforted by her mere presence,
remembering the journeys
that brought each of us here.

I think back to that bazaar
I think back to Mo—
his kindness, his authenticity.

A seemingly small thing, pink elephant
is my reminder of grace.

Jessi M. Atherton is a U.S. Army veteran of Operation Iraqi Freedom, who served in the Michigan Army National Guard from 2002-2006. In 2005, she deployed as a logistics specialist, with additional duties as a combat lifesaver. Atherton has worked in mental health as a registered nurse, including roles in veterans engagement, suicide prevention, and case management. She is a workshop facilitator for Warrior Writers, and a board member on the League of Minnesota Poets and the Minnesota Assistance Council for Veterans (MACV). She anticipates completing her degree as a Psychiatric Mental Health Nurse Practitioner at Walden University, St. Paul, Minnesota in 2024. She was named a 2023 Tillman Scholar by the Pat Tillman Foundation. Her debut poetry collection, The Time War Takes, *was published by Middle West Press in 2023.*

The Wooden Elephants of Herat
by Ben Weakley

I type *Afghanistan* into a search engine
that spits out words connected to places
and I get more places: *Kandahar, Khowst,*
Gardez, Herat.
      I never deployed to Herat.
But Herat is where a woodcarver cut
scraps of walnut into two elephants
I brought home from the war to give my son.
For eight years they roamed his room as he played
in the ivory carpet of his imagination
until the tusks, tiny as matchsticks, fell out.

He is ten now. He does not remember teething
on my dog tags or holding my sweat-stained
patrol cap in the Fort Knox gym the night I came home.
He does not remember stopping the car
to salute the flag when the trumpet played retreat
on post. He no longer plays with elephants,
and now I pack them into a cardboard box
with faded uniforms and dusty boots—
the relics we're unable to throw out
but no longer want to display.

Ben Weakley is a U.S. Army veteran of Iraq and Afghanistan, as well as a desk in the Pentagon. His debut poetry collection, HEAT + PRESSURE: Poems from War, *was published in 2022 by Middle West Press. He is a past poetry winner in the Col. Darron L. Wright Memorial Awards, as well as the Heroes Voices' Awards. He lives in Tennessee.*

Ode to My Skilcraft Pen
by Eric Chandler

I am in love with my Skilcraft ink pen.
Just government issue, for goodness sake.
Fancier ones? I lost probably ten.
My click-top, black ballpoint won't quit, won't break.
Humble in my flight suit pocket and then
it wrote miles of 9-lines without mistake.
More than a writing tool, it was my friend.
I left war, but didn't forget to take
my pen, worn smooth where the "Skilcraft" had been.

I fly for the airlines now and I take
notes with that same old trusty veteran.
Then I did something that made my heart break:
I left the ballpoint up on a shelf when
I ironed my shirt and forgot to take
it and put it in my pocket again.

I was in love with that Skilcraft ink pen.
More than a writing tool, it was my friend.
Remember that movie with Tom Hanks when,
at last, he sails away from the island,
but loses his battle buddy Wilson?
His best friend floats over the horizon?

I touched my pocket and felt my heart break.
No more war, but I'd forgotten to take
my old veteran pen, for goodness sake.
I've done far worse, but this guilt, I can't shake.

I was in love with that Skilcraft ink pen.

We fought wars together. It was my friend.
Like Tom Hanks wept when Wilson met his end,
I told my lost pen, "I'm sorry" again
and again
and again

Eric Chandler is the author of the poetry collections Kekekabic *(Finishing Line Press, 2022) and* Hugging This Rock: Poems of Earth & Sky, Love & War *(Middle West Press, 2017). He is a three-time poetry winner of the Col. Darron L. Wright Memorial Award, and a retired veteran of the U.S. Air Force and the Minnesota Air National Guard. Over Iraq and Afghanistan, he flew 145 combat missions in the F-16. He is happiest when on a Duluth, Minnesota trail with his wife, two children, and dog Leo.*

Something Else You Don't Need
by Liam Corley

Nudged awake by a loud enough
report, I pull near
the armor, helmet, and gun I know
the big mike will call for soon.

The Macedonians at the gate
survive, swaddled in checkpoints
manned by Afghan police, their history
not yet repeated.

Just Tuesday I walked past their HESCO barrier
and ducked around a concrete wall. Unaccountable,
the chicle boy I didn't buy from last week
holds out his hand again,
in it a pack of something
I don't need. Today he tumbled upward
like a leaf blown back to an awestruck
branch, tossed like my girl as she flies confiding
from my arms.

Only Tuesday he plucked a pen from my sleeve,
bold like so many urchins I've met in marketplaces or on
roadsides free of an overseeing eye.

Knowing how this type of news spreads,
I put the dime-store pen in a pocket, saving it
for some form to fill out, some check
in a box,

where it lies still in a bin,

buried between the drop-down
holster and boots that never fit
right, no good even
for this, the story of its life.

Liam Corley is a currently serving U.S. Navy reservist with deployment experiences in Iraq, Afghanistan, and throughout the Pacific area of operations. In his civilian career, Corley is a humanities professor at California State Polytechnic University, Pomona, Calif. His debut poetry collection, Unwound: Poems from Enduring War, *was published by Middle West Press in 2023.*

The Great Valentine's Barrier Riff
by David Abrams

A Valentine's Day memory from my war journal: Camp Buehring, Kuwait, 2005 (I was there a month before pushing north into Iraq.)

Camp Buehring (formerly "Camp Udairi"), Kuwait, 2005—Around midday, I had Specialist W. take a picture of me standing in front of one of the painted barriers here at this cold, dusty camp. I squinted against the sandblasting wind and tried to smile as she snapped the shutter. "One more, in case your eyes were closed," she said.

I didn't tell her my eyes are always closed when I think of Jean over here: I don't want to bring my wife to this war, even in my imagination, so I always pull my eyelids down like soundproof curtains over the helicopters, the rattle gunfire, the rumble of mortars—shutting it all out whenever I think of Jean back in Georgia. And, over here in this land of camels, grit, and graffiti concrete, I think about her all the time.

The painted barriers are something of a tradition here at Camp

Buehring. These are the inverted-T-shaped concrete blocks you see guiding traffic along construction areas or, lately on military installations, ringing a protective circles around important headquarters buildings.

Here at Camp Buehring, in addition to being used as traffic blockers/guiders, they also serve as canvases for military artists. As units rotate through here and elsewhere during Operation Iraqi Freedom, they have found illustrators (there's actually a Military Occupational Specialty in the Army for illustrators) or other talented Soldiers in the unit with time on their hands. The barriers are painted with the unit crests, logos, mascots, scenes from the war, or other symbols—sometimes, even cartoon depictions of their commanders. National Guard units often paint images that evoke their homes, such as state flags or well-celebrated scenes. A Wyoming unit, for example, painted their state's Devil's Tower National Monument.

As a collective monument themselves, the barriers leave a distinct and unique legacy at camp. Some of the barriers are getting old and weathered—I've even seen a couple here from way back in Operation Desert Storm, the first Gulf War—chipping and flaking away like DaVinci's "Last Supper" moldering in that cave ... But I digress.

In my snapshot, the barrier I am standing in front of is probably the most-photographed spot in all of Camp Buehring: On the left, it reads "From Camp Udari" (Buehring's name until this year). On the right, it reads "I love you." In the center is a big red heart. It's become quite popular in the weeks leading up to Valentine's Day. It seems like everybody had the same idea—take a picture of yourself in front of the painted barrier and send a personalized Valentine's Day picture to your sweetheart.

Jean seemed to like it and said she was going to use it as her computer desktop wallpaper background. Other than trading a few e-mails back and forth with my lover this morning, Valentine's Day was just another ordinary day in hell here in Kuwait.

The chow hall tried to make us feel special by serving prime rib, lobster, and shrimp tonight for dinner. They could have saved themselves the trouble because the "prime rib" turned out to be little more than a dried-out slab of beef (minus the usual au jus). Not even any horseradish sauce. I'd been warned by previous diners to stay away from the lobster, so I opted for the breaded shrimp. It went down the gullet as hard and dry as the beef. Oh well, at least there's always the bags of Cheetos they serve in the chow hall.

Specialist W. tells me her husband sent her an e-mail after he heard about this chow and said, "I hope you enjoyed your romantic dinner by camel-light." *Ha-ha-ha!*

David Abrams' retired in 2008 after a 20-year career as an active-duty U.S. Army journalist. He was named the Department of Defense's Military Journalist of the Year in 1994 and received several other military commendations throughout his career. His tours of duty took him to Thailand, Japan, Africa, Alaska, Texas, Georgia and The Pentagon. In 2005, he joined the 3rd Infantry Division and deployed to Baghdad in support of Operation Iraqi Freedom. His debut novel about the Iraq War, Fobbit, *was named a New York Times Notable Book of 2012 and a Best Book of 2012 by Paste Magazine, St. Louis Post-Dispatch, and Barnes and Noble. A second war novel,* Brave Deeds, *followed in 2016. Abrams was born in Pennsylvania and grew up in Jackson, Wyoming. He holds an undergraduate degree in English from the University of Oregon and a Master of Fine Arts in Creative Writing from the University of Alaska-Fairbanks. He now lives in Montana with his wife.*

Two Rugs, Two Wars
by Dale Eikmeier

An Afghan war rug hangs on my office wall, a memento of the Global War on Terrorism. An Iraqi war rug lies on the floor, a memento of a different war. One is primitive folk art and genuine. The other is just souvenir kitsch.

Afghans have a long history of war and an even longer history of rug-making, so it's natural they combined the two by making "War Rugs." My example shows a map of Afghanistan, along with Soviet tanks, artillery, helicopters, and the ubiquitous Kalashnikov rifle. Its colors are subdued and natural; they are easy on the eye. The wool weave is tight and feels good. I imagine a young boy spent days at a loom weaving it under the careful eye of a master weaver.

The Iraqis also have a long history of war, but not rug-making. But they know how and when to make a buck, which is how my second war rug came to be. My Iraqi rug is a low-quality imported bathmat, probably factory made in Pakistan. The motif includes machine-embroidered American and Iraqi flags, a camel, and the words "Operation Iraqi Freedom." It's a combination of Leprechaun and baby-poop green and feels synthetic. Using an embroidery machine, its "maker" probably spent minutes on it.

One of my rugs could hang in a folk-art gallery, while the other would best be displayed in a basement next to a velvet Elvis painting.

In the currency of memory, however, they are equal.

Dale Eikmeier received a commission as after graduating San Jose State University in 1978. His 30-year career included assignments in Germany, Bosnia, Korea, and Iraq. Assignments included an Air Defense Artillery battery command in the 7th Infantry Division, a battalion operations officer position in the 1st Infantry Division, and a battalion command in the 101st Airborne Division. He is a veteran of the First Gulf War and the Iraq War.

dust bunnies and combat boots
by Randy Brown

The not-a-prayer rug
beside my Army bunk
guards bare soles at night.

Pictograms of Kalashnikovs,
grenades, and A.P.C.s
are part of its bazaar tapestry.

DFAC peanut butter
laces the spring I have
set under my cot, next to my boots.

The snap of the trap
summons me to my prey.
I find each of us, kneeling.

In the dark.
In a box.
Facing west.

Randy Brown embedded with his former Iowa Army National Guard unit as a civilian journalist in Afghanistan, May-June 2011. A 20-year veteran with a previous overseas deployment, he subsequently authored the poetry collection Welcome to FOB Haiku: War Poems from Inside the Wire (2015). *He is the author or co-editor of multiple books and anthologies, including* Why We Write: Craft Essays on Writing War. *On social media, he often still writes about military themes, topics, and literature under his original mil-blogging pseudonym "Charlie Sherpa."*

War Trophies
by Nathan Webster

On my basement wall hangs a poster-sized, hand-drawn schematic of a fighter plane cockpit. The poster is a teaching diagram, words and arrows pointing at and describing a joystick and controls. On March 3, 1991, I carved it from its original frame in a classroom at Iraq's Tallil Air Force Base, then shepherded it home. Over 30 years, the poster was creased, folded, faded, tacked up on other walls, now re-framed and matted for proper presentation. It migrated, from displays of dramatic visibility, down to fill the basement's empty space.

The poster is my war trophy, my aggressive act of liberation. Not passive nostalgia, like my combat boots, issued brand new, before sand scuffed the leather, sweat stained the canvas. No foot ached in them but mine. The boots became bookends, bracketing that story, beginning to end.

For my poster, my war trophy, I arrived in the story's middle.

The opposite of a war trophy is not absence, it's interruption.

Iraqi airmen returned to post-war Tallil to face an impossible clean-up: broken furniture, ransacked offices, bombed hangers and cratered runways.

My poster's artist would have stood in the classroom. He would trace his fingers inside the empty frame, along the knife-cut edges that I left behind. His time, choosing ink and colors, drawing objects and words, for a pilot's lesson, for his portfolio. Those locusts, he would say, those locust motherfuckers. A friend would come up and slap him on the back. You've gone worldwide, the friend would tell him, somebody's hanging it somewhere. They would laugh, as soldiers do.

Would the artist have even cared? Fair question.

In my poster's lower left corner is penciled Arabic script, enough tiny handwritten letters to prove he cared, like my sweat stained my boots. His name was M. Waleed.

Nathan Webster is a U.S. Army veteran of Operation Desert Storm, and also reported from Iraq several times from 2007 to 2009. He currently works in higher education. He work from Iraq was published in dozens of stateside newspapers. He contributes periodic reviews to The Daily Beast and other venues.

Your Future Love
by Callie S. Blackstone

Picture the map: flat
on the wall of your grade
school classroom. Picture
the solid blue background, pierced
by triangle waves—
the solid masses of color
that represent continents, countries. Trace
your fingers around the compass, violet
and red borders. Find the border
outlining the country
you have learned is called Iraq. He is there,
in the future. He is there,
and he is holding a gun.
You do not know
who he is. Who is the man
who easily wears the costume, the camouflage,
that even you in the second grade understand
equates death? The blackness
of his shined boots signifies extermination.
He is the end. The end
of the map, the unexplored,
the unobtainable. You will never reach
him, your fingers will fall off the slick paper,
and you will find yourself groping for him
in the middle of the night for
the rest of your life.

Callie S. Blackstone writes both poetry and prose. Her debut chapbook sing eternal *is available through Bottlecap Press. Visit: calliesblackstone.com.*

Guacamole
by Jessi M. Atherton

You make the best guacamole—
that's what everyone said.
It was a party favorite,
just like you.
But we all know some favorites don't last.

You wouldn't give me the recipe
so, I watched
when you weren't paying attention:

3 ripe avocadoes,
half the juice from a lime.
You forgot to mention you had a daughter
with your high-school girlfriend.

A few dashes of garlic powder.
You lied about the meaning of the tattoo of your forearm.
22 was not the number of soldiers you lost on your first deployment.
It was the gang number given to you.
You were given the choice:
Go to jail or join the military.
We met in the Army.

Salt and pepper to taste, but not too much.
Like the time we went to Corpus Christi and swam in the ocean
You accused me, of cheating, out of the blue—
pulled my hair during the argument.

A sprinkle of cayenne powder for a little spice.
You punched a hole in the wall
right next to my head when

I told you I was planning to leave.
You made me promise not to call the MPs.

A few dashes of onion powder.
Chopped tomatoes—Roma was your favorite.
Your mother called me *mia.*
I heard you tell your cousin that it was great
that I was naive and young.

I watched. I saw
a broken soldier with deep trauma
who didn't love himself, and
could never love me.
You cried when I left.

I've perfected the guacamole recipe.
Everyone loves it:
Best served with tortilla chips and
a strong drink.

Jessi M. Atherton is a U.S. Army veteran of Operation Iraqi Freedom, who served in the Michigan Army National Guard from 2002-2006. In 2005, she deployed as a logistics specialist, with additional duties as a combat lifesaver. Atherton has worked in mental health as a registered nurse, including roles in veterans engagement, suicide prevention, and case management. She is a facilitator for Warrior Writers, and a board member on the League of Minnesota Poets and the Minnesota Assistance Council for Veterans (MACV). She anticipates completing her degree as a Psychiatric Mental Health Nurse Practitioner at Walden University, St. Paul, Minnesota in 2024. She is the recipient of a 2019 Minnesota Humanities Veterans' Voices award, and was recently named a 2023 Tillman Scholar by the Pat Tillman Foundation. Her debut poetry collection, The Time War Takes, *was published by Middle West Press in 2023.*

Pistol
by Jon Alston

Cold worn steel holstered in dilapidated leather. Near a century ago. Seen battle. Many battles; long gouges along the barrel, dents and pocks of being dropped on the grip, the trigger polished from oily skin. Sixty years ago a German soldier died wearing the gun around his waist; Captain perhaps, or maybe a General. He's dead. Now the gun is on my table. Chambers empty. Gone cold for decades in a box lost to a dying garage. But now it's here, still cold, sealed behind the raw chapped leather. Unlike days before, when people shot at close range, watched each other die, made sure you knew who killed you, what the faces of the living and the dead do, how your body smells when the spirit leaves; what the dusty airs tastes like. This pistol will do nothing, though. It will be examined. People will be impressed by its age, its importance for history, for family. It will pass through generations. But it will do nothing.

Jon Alston is a husband, father, writer, book maker, and artist. With a graduate degree in creative writing, he teaches a variety of subjects at a homeschool learning center co-op, and focuses all his free energy on his own writing and handmade books. He has been published in more than 30 literary journals, including Conium Review, The Binnacle, The Metric, The Machinery, *and appeared in the collaborative artist project The Encyclopedia Project. Both of his grandfathers were in World War II.*

Battlefield Typewriter
by Martin Ott

When I read that J.C. Salinger interrogated
POWs in Germany, and knocked back stiff
ones with Hemingway, it explained everything
and nothing. The ashes of Dachau cascaded
off the mop tops of his dead-eyed youngsters,
bone-dry dandruff, hidden but omnipresent.
Confessions filled his battlefield typewriter
and keys clacked like gold fillings against
wet pavement, his fingertips callused from
the impossible gravity. He fell apart once
and forever disdained to explain the lost
ones, the tracks a distant killer makes
before disappearing onto a folded page.

*Martin Ott is the author of ten books of poetry and fiction, the former
including* Captive *(C&R Press, 2011);* Underdays *(University of Notre
Dame Press, 2015); and* Lessons in Camouflage *(C&R Press, 2018) A
former U.S. Army interrogator and longtime resident of Southern
California, Ott works as a communications professional and develops for
TV and film in-between other writing projects. His latest novel,* Shadow
Dance, *is a noirish narrative about a U.S. veteran who gets caught up in
the dark underworld of an Iranian-run gentlemen's club in Los Angeles.*

The Shield
by Peyton Roberts

The tail-end of moving is underway, and we are unpacking boxes at our new home. Across the living room, my husband pulls the packing paper off a heavy wooden plaque.

The plaque was a gift from his command a few tours ago. Etched into the grain is the command's emblem, a black and gold shield.

My husband inspects the plaque, pleased it has arrived intact. But my chest tightens. Tears sneak into my eyes remembering that deployment.

He left on a Monday in October. Our daughter was barely potty trained. Our baby son could hardly lift his head. I didn't yet have my footing as a mother of two. I didn't know my strengths or my limits.

As autumn days shortened into winter darkness, my life solo parenting our little ones rotated on a carousel of feedings and tantrums and sleeplessness, long days bleeding into longer nights.

Throughout it all, the cloak of responsibility engulfed me. One slip up. One tiny mistake could be tragic. I wondered if one of us wouldn't make it to homecoming.

My role, I determined, was to get us there. I prayed he'd be there, too.

On a chilly day in March, my husband came home. But not every husband did. The joy of that reunion forever rests at half-mast.

Now in the living room at our new address, my husband asks, "Where should we hang this?"

He positions the heavy plaque against a prominent wall near the fireplace. "How about here?"

Showcasing this dark reminder makes me cringe. But I can hear in my husband's voice that, to him, this plaque means something entirely different.

I don't smile, but I nod.

He reaches in his pocket for a nail.

I swallow the lump in my throat and hand him a hammer.

Peyton Roberts is the author of Beneath the Seams *(Scrivenings Press, 2021), as well as a seasoned Navy spouse and mom. Her stories about military family life have appeared in The New York Times, Modern Love Podcast and more. Her forthcoming book,* My Dearest Bea, *is an intimate collection of love letters her grandfather wrote to her grandmother aboard the USS Midway. Visit: www.peyton-roberts.com*

Snapshot
by Layle Keane Chambers

my lockscreen
you
suited up
strapped into
cockpit, T-6 Texan
masked and tubed to fight
first solo flight
$ ride

the morning
strums
face fear
practice
Like sit-ups
Like a fist
beneath belly flesh
iron

Layle Keane Chambers is a teaching artist currently living in Folly Beach, South Carolina. She received her undergraduate fine arts degree from Ithaca College and master's degree from University of Texas at El Paso in Theatre Arts. She is a professor emeritus of theatre at Doña Community College, Las Cruces, New Mexico; a Master of Fine Arts candidate in creative writing at New Mexico State University; and a proud U.S. Air Force mom.

Half-Life
by Lee Anne Gallaway-Mitchell

My husband keeps an eye on the dumpster outside the squadron.

At our first base, he finds the top of an ejection seat from an aircraft and brings it home just as the movers are about to arrive for our PCS. The packers give the ejection seat its own box. We discover it months later, the box labeled, "Objection seat." *I object!* becomes a punchline for punching out. It's always metaphorical, but not always funny.

After his first deployment, he takes a carpentry class, which is where he spends most of his two weeks leave. He makes a wine rack. In the years since, he has used his woodworking skills to fashion a hatrack with 30mm cartridge cases as hangers. He mounts old gun barrels from the jet onto large wooden plaques for recently retired pilots. I wonder if he retreats to make these objects when work starts taking too much from him. Whatever he makes, he gives. We live in Tucson, Arizona, home of the Boneyard, where a few thousand planes lounge on 2,600 acres, just waiting to get scavenged for parts.

But what does he carry in his body? What hitched a ride home from the Bagram burn pits, the carcinogens from his jet, from all the places we have lived with varying levels of pollution? We stayed inside for weeks when the Yellow Dust traveled across Korea.

We are told to have proof we lived in Korea during the worst of these seasonal dust storms just in case we have lung problems later. As our son would have said at that time, age four, *Just be case.*

Also, our son: *It won't be for-long-ever, just for-little-ever.*

Lee Anne Gallaway-Mitchell lives and writes in Tucson, Arizon. She holds an Master of Fine Arts in creative writing from the University of Arizona. Her essays have won The Florida Review Editor's Award, the Arts & Letters Susan Atefat Prize for Creative Nonfiction, and Boulevard Magazine's Nonfiction Contest for Emerging Writers. She is at work on an essay collection, Campfollowers.

Black Bracelets
by Mason Rodrigue

You hit the fleet as a stupid boot and you
wanna be salty like seniors.
You want long hair and
tattoos and
Caffeine dependences and
Nicotine addiction and
Black bracelets
Because they have all that
until
You get all that.

Mason Rodrigue (pronounced "road-rig") was born in southern Louisiana. The terrorist attacks of September 11, 2001 led him to enlist in the U.S. Marine Corps and eventually his service in Syria as a machine-gunner. Rodrigue is the author Rock Eater *(Dead Reckoning Collective, 2023), a collection of poetry and philosophy about "what happens when a man gets turned into a weapon and aimed at nothing."*

Jewelry
by Abby E. Murray

I am bribed with jewelry to behave.
My wrists clink against the table.
My fingers catch on the backs of chairs
and my neck glitters with guilt.
A white gold bracelet of braided light
lures me to another military ball
where Captain Stanley hacks the cork
from a champagne bottle with his saber
and accidentally fills his wife's forearm
with shattered bits of glass.
I chase the emerald pieces
as if I've earned them, hustle Mrs. Stanley
to the bathroom and pinch each gem
with the silver tweezers of my Swiss Army knife,
patch her up with Band-Aids from my wallet.
This is how I remain calm.
This is how I lodge a complaint.
I rub despair into the amethyst on my knuckle,
name every shade of green in a pearl necklace
and pray the rosary into a Swarovski knot:
one shiny rock for every night the phone doesn't ring,
every close call, every care package,
a clasp on the end of every year.
My initials shine in lapis lazuli,
my pinky is wrapped in platinum.
I want to protest the diamond band
you order from Afghanistan, the heft
and glare of each stone that says
there is no more, your body is the final prize,
carved from the earth and polished.

Abby E. Murray was the 2019-2021 poet laureate for the City of Tacoma, Washington. Her first book, Hail and Farewell, *won the 2019 Perugia Press Poetry Prize and was a finalist for the Washington State Book Award. Previous chapbooks include* How to Be Married after Iraq *(Finishing Line Press, 2018),* Quick Draw: Poems from a Soldier's Wife *(Finishing Line Press, 2012), and* Me and Coyote *(Lost Horse Books, 2010). She is the founding editor of* Collateral, *a literary journal publishing work concerned with the impact of violent conflict and military service beyond the combat zone. She teaches rhetoric in writing military strategy for U.S. Army War College fellows at the University of Washington, as well as poetry workshops at community centers, coffee shops, military posts, detention centers, shelters, and schools.*

Jesus Hold My Earring
by Nancy Brown

Homecoming
Is late
And in the Dark,
On a Cold Night,
But I am wearing
The Dress
Anyway
Because I tried on
Sixteen
Which is more than
I did for my
Wedding.
I have made a
Dinner
And leave the table
Set
For a Family of Five
Instead of plastic plates,
Sippy cups,
And Scraps
Of the Last Seven Months.
The boys are dressed
"Subtly Matching"
As though prepared for photos
That we will not take.
Because we are shaking,
And laughing.
And smiling.
Simultaneously
Hurting,
And Healing.
I have not cried

In Seven Months.
It was never safe to cry.
On the way to the airport,
To the end of the deployment,
I realize
I'm missing an earring
That I picked out
Especially
For This Night.
And in the Darkness,
I feel all the un-wept tears
Come rolling down my face.
So that when we
(Which still means,
The Children and I)
Finally see him,
Grab him,
Hug him,
Smile at him,
Pull him to the car ...
My make-up camouflage
Has dissolved
And drenched the front of the
Sixteenth Dress
And I am left facing him
Bare,
With one little earring
Dangling by a thread.

Nancy Brown is a U.S. Marine Corps spouse who writes narrative verse. She mothers three boys, and works as an educator. She believes in the power of story to unite us all and make us feel less alone.

Gift
by Yvette Viets Flaten

The Yank was visiting his English girlfriend in mid-May 1944, and Alan Ette, the girl's brother, arrived home on a surprise leave. Both men knew that something was up—something was coming—the long-awaited strike against Nazi Germany.

Tea finished, Alan got up to head back to barracks. The Yank, in a moment of generosity, handed the Limey a pack of his cigarettes. "Tah," Alan said. "Thanks, I shall keep these for special ..."

At 9:00 on D-Day, Alan came ashore on Juno Beach, landing on French soil at Courseulles-sûr-Mer. Two weeks later, clambering over a hedgerow, he saw the flash of the German's weapon. He fell backward, his neck on fire, his helmet rolling away, filled with blood. Later, a Royal Army Medical Corps orderly knelt down, cutting away his tunic. The orderly mumbled something, put Alan's cigarettes on his chest, and left him lying under an apple tree, paralyzed.

"Hey! Over here!" someone shouted. "Here's one of ours!" Alan heard the voices but he couldn't respond. He was fading away.

When he came to, he was in a hospital bed. A woman in white walked by. "Sister ... Sister ... " Alan whispered.

"Who do you think you're calling 'sister?!'" the U.S. Army Nurse snapped, angry that a soldier was being forward—here, of all places!

"Sister ... where am I?"

The nurse stopped short. She recognized an English accent. To him, sister meant nurse. "You're on a hospital ship in the Channel. You've just had surgery. But how ...?"

How? The Royal Army Medical Corpsman had placed the Yank's spontaneous gift on Alan's chest. The next medics through the orchard were American. They recognized the cigarettes and tagged him for evacuation, assuming Alan was one of their own.

And that chance gift? That cigarette packet? *Lucky Strikes!* The luckiest strike of them all.

Yvette Viets Flaten grew up in a U.S. Air Force family, living overseas in England, France, and Spain, and stateside in Nevada, North Dakota, and Washington state. Of her family's various military experiences, Flaten notes: "One English uncle was in North Africa in World War II, and fought against Rommel. The other, who is the center of my story, was in the Durham Light Infantry and went ashore on Juno Beach on D-Day. The story is true. My father gave him the Lucky Strikes. They later became brothers-in-law."

forty-three years and counting
by Paul Hellweg

Invitations declined,
alone by choice
Christmas eve,
fiber-optic tree twinkling
green yellow red blue,
indulging in Cabernet Sauvignon
and Double Black Scotch,
watching the History Channel's
Vietnam in HD.
Disc Two: *An Endless War,*
1968-1969, my time,
year of escalation,
dead bodies twisted and strewn
like discarded wrapping paper,
blood spilled far and near,
my own a part of the sacrifice.

Read poem today
about how an injured animal
will isolate itself, finding quiet place
to lick its wounds
before rejoining the pack.
Poem didn't say
how long.

Paul Hellweg is a U.S. Army veteran and freelance writer. His poetry has been nominated for the Pushcart Prize and the Best of the Net Awards. His first collection, Ode to a Drunken Muse, *was published by Alien Buddha Press. A collection of Vietnam War poems,* When Eagles Vie with Valkyries, *is forthcoming from Main Street Rag Press.*

Life and The First Gulf War

by Juan Manuel Pérez

Sonnet No. 43

Thirty years ago, music was my peace those happy Walkman cassette
    playing days
I took some favorite tapes to war with me
like Megadeth's "Peace Sells But Whose Buying"
as well as their new album, "Rust In Peace"
also with me, Ozzy's "Bark At The Moon"
when not on duty, I read, wrote, listened
however, the only song that brought me
back home, far away from that place of war
was not even heavy metal at all
it was Greenwood's "God Bless The USA"
it was on the plane coming home with me
it was everywhere back in that decade
... how'd this song become a racist anthem?

Juan Manuel Pérez, a Mexican-American poet of Indigenous descent and the Poet Laureate for Corpus Christi, Texas (2019-2020), is the author of numerous poetry books including Thirty Years Ago: Life and the First Gulf War *(The House of the Fighting Chupacabras Press, 2023). Juan, a former migrant worker, is also the 2021 Horror Authors Guild's Inaugural Lifetime Achievement Award winner and a recipient of a 2021 Horror Writers Association Diversity Grant. To learn more about this award-winning poet, combat veteran, teacher, and gourd dancer, visit: www.juanmperez.com*

Still Carried
by Benjamin B. White

A grenade pin,
dog tags,
a P-38,
the fatigues,
a wooden cross from the Chaplain—
          all these takeaways
augment the memories
that carry me
while remembering
          the baby-eating Soviets,
the appreciation
for the host country,
          the complicated simplicity
          of Cold War duty,
and the beauty of youth
caught in a constant truth
          so often triggered
by the smell
of diesel exhaust
that always takes me back
to the tracks and trucks
          in a mobilized convoy
rolling out
in a morning muster
to flex American posture
          for the foreign diplomats
          we knew were watching

Benjamin B. White was pursuing a career in baseball when a worn-out shoulder diverted his life's course. After an enlistment in the U.S. Army, White ended up as an officer in the U.S. Coast Guard. He served 22 years. Now retired, the poet and novelist lives near Albuquerque, New Mexico. He is the author of more than eight books, including Always Ready: Poems from a Life in the U. S. Coast Guard *(Middle West Press, 2022).*

The Leatherman
by Peter Molin

When I first joined the Army I noticed that many soldiers more experienced than me carried on their belt not just a jackknife, but a particular kind of multi-purpose tool called a Leatherman. The Leatherman resembled a Swiss Army Knife, but without the elegance of design. Where a bright Swiss Army knife seems as if it could have been made by Swiss artisans, a Leatherman was dull black and seemed forged out of cheap or leftover tin. It wasn't even all that functional. When I got my hands on one for the first time, I noticed that the blade was neither long nor sharp, the bottle- and can-openers marginally useful, and the scissors and saw functions pathetic. A saw? The only function that seemed useful were the pliers, but how often was that necessary? Plus, when I priced a Leatherman in the local military gear store, it seemed very expensive for what you were getting.

But that's the thing—the idea was not to *buy* a Leatherman with your own money, but to *obtain* one through your unit supply shop. Leathermans were cool; the soldiers who had them whipped them out with panache, and were all the time finding some little task to do that could only be performed with one of the multitools. And not only did all the cool guys have a Leatherman, they were able to obtain them for *free*, because they knew someone in supply with whom they had made a deal to get one off-the-books. To actually have to buy a Leatherman was evidence that you weren't yet worthy enough to wield one. If you were a newbie in the unit, not having a Leatherman was a sign of exactly how new you were.

And so it was for the first twenty years of my military career. No Leatherman for me, just ordinary old pocket-knives of one brand or another. But then, in training at Fort Riley, Kansas prior to deployment to Afghanistan, we drew a lot of personal gear. In fact, we drew gear three times at three different places, and there were individual-issue items as well. And every time we opened our bag to receive new equipment, the supply guy would drop in a Leatherman. Not once, not twice, not three times. By the time I packed my duffel bags to fly to Afghanistan I had four Leathermans.

I didn't think I was now cool, but something had changed. Things were different.

Peter Molin is a retired U.S. Army officer who served in the 2nd Infantry, 10th Mountain, and 82nd Airborne divisions. In addition to deployments to the Sinai, Kosovo, and Afghanistan, he taught in the Department of English and Philosophy at the United States Military Academy, West Point, New York. His literary, film, and cultural criticism have appeared in such venues as The Wrath-Bearing Tree *and his own Time Now blog.*

Trench Whistle
by Jehanne Dubrow

I call my dogs with it, a blast of sound
that summons them from far across the grass,
and they come running at the trill, each hound
a blur of lead-gray fur, a yipping mass.
They only know the pleasures of the lawn,
the clovered fragrances. This cry I make
is like a strident bird that's quickly gone,
the branch now bare, the air that seems to shake
with what was there before. They can't conceive:
this whistle mobilized whole companies
of men. Its voice was sharp enough to cleave
the clamor of the front, the batteries
of mortar rounds. I call, and each dog runs,
unhunted by the terror of the guns.

Jehanne Dubrow is the author of nine poetry collections, including most recently Wild Kingdom *(Louisiana State University Press, 2021), and two books of creative nonfiction,* throughsmoke: an essay in notes *(New Rivers Press, 2019) and* Taste: A Book of Small Bites *(Columbia University Press, 2022). Her third book of nonfiction,* Exhibitions: Essays On Art & Atrocity, *was published by University of New Mexico Press in 2023. Her writing has appeared in* POETRY, New England Review, Colorado Review, *and* The Southern Review. *She is a professor of creative writing at the University of North Texas.*

Wedding Ring
by *Chad Corrigan*

He had survived his third combat tour
But his marriage hadn't.
He took the ring off
And chucked it
outside the wire
Into the Afghan sand
He was never coming back to get it
He liked the finality of that.

Chad Corrigan is a U.S. Army officer and helicopter pilot. He holds an undergraduate degree in philosophy and political science from Stonehill College, Easton, Massachusetts. He also holds a graduate degree in public policy and management from the University of Pittsburgh. He is a former professor of military science and a department chair at Boston University, and a fellow at Harvard University. His writing has appeared in the Why We Write: Craft Essays on Writing War, *and in the journals* From the Green Notebook, The Wrath-Bearing Tree, *and* As You Were. *He has served in Iraq, Afghanistan, Syria, South Korea, and Germany, as well as at the Pentagon. He is a member of the Military Writers Guild.*

Hippie Car Stickers (haiku)
by Andria Williams

When we moved on-base
I pulled off my car stickers
One by one by one.

Andria Williams is the author of The Longest Night *(Random House, 2016), a novel set among U.S. Army families in Idaho in the late 1950s. It was chosen as Amazon's "Debut Novel of the Month" for January 2016. A new novel,* The Waiting World, *is forthcoming from MilSpeak Books. She holds an undergraduate degree in English from the University of California, Berkeley, and a Master of Fine Arts in Creative Writing at University of Minnesota. Williams is a founding editor of the* Military Spouse Book Review, *and a former editor-in-chief at* The Wrath-Bearing Tree. *She and her family live near Colorado Springs, Colorado, where her active-duty U.S. Navy husband is stationed.*

Weapons, Vehicles & Equipment

No Beer on the Drop Zone
by Nancy Stroer

My hand touches the tattered copy of my civilian jump log in the bottom of a box and immediately I'm in Lahr, Germany in 1990. I'm lying on my back, head resting on the small of the back of The Ranger who's lying on his stomach, studying the grass. We're always perpendicular to each other, but in that moment my joy warms me like the first real day of Spring. It cools me like the first swig of a brown-bottled Pilsner. We're waiting for our turn to jump from the Cessna 182 that the Black Forest Parachute Club calls the Yellow Banana.

"Ew." My old roommate had this way of wrinkling her nose when displeased. "Why Germany when you could go to the 82nd Airborne?" She was on her way to Fayettenam. By choice.

Why, indeed, choose a hellhole riddled with fire ant hills, strip clubs and strip malls, when you could do your jumps with a bunch of off-duty Canadian zoomies whose only rule was no beer on the drop zone? Beer before the drop zone? Sure. Right after you get off? You betcha. Just not on the actual drop zone.

You had to pack your own chute, but ...

Details, details, I say, cracking open another cold one and spreading the rectangular sheet of nylon flat in the sun, beginning to make my folds, ready to go again. The Ranger looks concerned. He'd been the one shaking in the Banana while I stepped out onto the struts, falling backward, soft in the sky. Trusting the Jump Master and the drogue chute to do their jobs. Three-thousand feet affords plenty of perspective, and at that elevation, life looked damn near perfect. Only off-duty, though, only in the sunshine. But on or off the drop zone, buzzing in love.

Nancy Stroer arrived in Germany as a newly commissioned U.S. Army maintenance officer two days after the fall of the Berlin Wall. She took a

European separation after her first-duty station so she could stay overseas, and has lived and worked in military communities in Germany, Turkey, Japan, and the United Kingdom. Her writing has appeared in Stars and Stripes *newspaper and* Soldiers *magazine, as well as the* Hallaren Lit Mag *and* The Wrath-Bearing Tree. *She is currently working on a series of linked novels about women in the U.S. Army, not one word of which is true.*

Wedding Arch
by Lisa Stice

Kiss for Permission to pass underneath
etched Mamaluke blades raised and lowered
and raised again

Pass, uniformed in tradition,
between lines of men
who cannot kiss their wives this day.

Press his cold medals against
your collar bone and
smile though he will be gone

nine months from now.
Wish you could have held
that kiss a little longer

before the final blade
strikes your backside
and you are formally initiated.

Lisa Stice is the author of three full-length poetry collections, FORCES *(Middle West Press, 2021),* Permanent Change of Station *(Middle West Press, 2018) and* Uniform *(Aldrich Press, 2016), and a chapbook,* Desert *(Prolific Press, 2018). She currently serves as poetry editor for* Inklette Magazine, *as well as the Military Spouse Book Review. She writes for the Military Spouse Fine Artists Network (Milspo-FAN), and is an associate editor for Middle West Press. She currently lives in North Carolina with her U.S. Marine husband, daughter, and dog Seamus.*

Last Four
by Jessi M. Atherton

In the plane, on the way
to Iraq
      I got a black Sharpie

handed back to me from
a person in the row
         ahead of me.

I was instructed to write
           the last four

of my social security number
on the heels of my boots
in case my legs ever got
        separated

from the rest of my body.

When I left country
I didn't get another marker—
even though by then
        my heart

was a part of my body

no longer
    connected

Jessi M. Atherton is a U.S. Army veteran of Operation Iraqi Freedom, who served in the Michigan Army National Guard from 2002-2006. In 2005, she deployed as a logistics specialist, with additional duties as a combat lifesaver. Atherton has worked in mental health as a registered nurse, including roles in veterans engagement, suicide prevention, and case management. She is a facilitator for Warrior Writers, and a board member on the League of Minnesota Poets and the Minnesota Assistance Council for Veterans (MACV). She anticipates completing her degree as a Psychiatric Mental Health Nurse Practitioner at Walden University, St. Paul, Minnesota in 2024. She is the recipient of a 2019 Minnesota Humanities Veterans' Voices award, and was recently named a 2023 Tillman Scholar by the Pat Tillman Foundation. Her debut poetry collection, The Time War Takes, *was published by Middle West Press in 2023.*

Holding Maps
by Gerardo "Tony" Mena

I remember the day we learned that our platoon commander
    had no idea
how to read

a map. We followed the streets around in a circle, twice,
    on a road notorious
for buried IEDs.

I developed my first mantra that day. I couldn't stop whispering it.
    It was as simple
as our snatch and grab

mission should have been. As simple as looking at a map. My mouth
    formed shapes
that repeated: Not my legs.

Gerardo "Tony" Mena is an Operation Iraqi Freedom veteran who spent six years in the Special Operations community with the Reconnaissance Marines as a Special Amphibious Reconnaissance Corpsman (SARC). During his tour in Iraq, he was awarded a Navy Achievement Medal with a "V" device for multiple acts of bravery. His debut poetry collection, The Shape of Our Faces No Longer Matters, *was published by Southeast Missouri State University Press in 2014.*

Soldier's Song
by Ben Weakley

More life exists in the tip
of a bullet smacking
the concrete wall
beside your head

than in a decade spent
commuting to work in traffic
paying the mortgage on time
loving one woman and two children
and taking vacations at the beach.

More sweat pours
more breath gasps and heaves
more heartbeats pulsate
in the intimate space

between shock wave and steel debris
than you could ever find
in any lover's fingers.

More enlightenment
flows through the dust cloud
rising from broken asphalt
to drag you underneath
the opened ground

than in a hundred years
of anything else in this life
or the next.

Ben Weakley is a U.S. Army veteran of Iraq and Afghanistan, as well as a desk in the Pentagon. His debut poetry collection, HEAT + PRESSURE: Poems from War, *was published in 2022 by Middle West Press. He is a past poetry winner in the Col. Darron L. Wright Memorial Awards, as well as the Heroes Voices' Awards. He lives in Tennessee.*

Sextant
by Benjamin B. White

The requirement was to use a sextant
And get a reading of a heading
Within so many degrees—
                    So the Officer Candidate
Went through the prescribed steps,
Called out the heading,
And hid his surprise
When it was checked off
                    As correct.

Proudly relieved
To be sextant-qualified,
He subsequently
Put the instrument down
And walked away
Never to apply
                    The skill
            Ever again

Benjamin B. White was pursuing a career in baseball when a worn-out shoulder diverted his life's course. After an enlistment in the U.S. Army, White ended up as an officer in the U.S. Coast Guard. He served 22 years. Now retired, the poet and novelist lives near Albuquerque, New Mexico. He is the author of more than eight books, including Always Ready: Poems from a Life in the U. S. Coast Guard *(Middle West Press, 2022).*

Man-Machine
by Eric Chandler

The pilot who wrote *The Little Prince* said
we take the airplane for granted.
Because of its modern reliability,

"... the machine does not isolate man
from the great problems of nature but
plunges him more deeply into them."

In high school, my buddy Mel
had a thought about
the pads that football players wear:

They protect you so well that
if you even notice you've been hit,
it's already
really
really
bad.

One flier always shakes the crew chief's hand
before closing the bubble of glass and
fusing himself to the fuselage:

It might be
the last time
I touch
a human being

Eric Chandler is the author of the poetry collections Kekekabic *(Finishing Line Press, 2022) and* Hugging This Rock: Poems of Earth & Sky, Love & War *(Middle West Press, 2017). He is a three-time poetry winner of the Col. Darron L. Wright Memorial Award, and a retired veteran of the U.S. Air Force and the Minnesota Air National Guard. Over Iraq and Afghanistan, he flew 145 combat missions in the F-16. He is happiest when on a Duluth, Minnesota trail with his wife, two children, and dog Leo.*

Enlightenment
by Dennis Maulsby

> *"The annual rainfall is heavy in all regions and torrential in many.*
> *It is heaviest at Hue, which has an annual average of 128 inches."*
> *—U.S. Army Handbook for Vietnam, 1962 edition*

The army chopper cycles in,
churns the heavy, dirty air,
buffets sweat-faced men and women.

Loaded, it climbs to cool clean heights.
Below, curled rivers and layered jungle canopy,
blues and greens in ruffled Chinese silks.

Above, creamy monsoon clouds
release slanted gray pillars of rain,
elephant legs walking the land.

Sunlight flits across the mist,
millions of tiny prisms refract color
displaying spread peacock feathers.

An aching beauty possesses us.
God whispers to our becalmed souls.
We are only short-timers here.

Dennis Maulsby is a retired bank president and U.S. Army veteran living in Ames, Iowa. He is the author of an award-winning book of war poetry, Near Death/Near Life (2015), *a book of short stories,* Free Fire Zone (2016), *as well as others. Visit: www.dennismaulsby.com*

Life and The First Gulf War
by Juan Manuel Pérez

Sonnet No. 45

Thirty years ago, flying to the front
early in the war, before the big push
the transport helicopter lifted up
into the bluest skies I'd seen that month
two of our squads onboard "Marine Airlines"
gunner by the door manning the big one
looking at the ground; beautiful, quiet
despite the immediate chopper noise
caught in the moment of HIS creation
the back opening view was picturesque
... until the fear of a volatile sky
infected me with an old memory
I was a child with a new pellet gun
shooting at birds as they flew through the air

Juan Manuel Pérez, a Mexican-American poet of Indigenous descent and the Poet Laureate for Corpus Christi, Texas (2019-2020), is the author of numerous poetry books including Thirty Years Ago: Life and the First Gulf War *(The House of the Fighting Chupacabras Press, 2023). Juan, a former migrant worker, is also the 2021 Horror Authors Guild's Inaugural Lifetime Achievement Award winner and a recipient of a 2021 Horror Writers Association Diversity Grant. To learn more about this award-winning poet, combat veteran, teacher, and gourd dancer, visit: www.juanmperez.com*

The Unholy Cabal
by GOODW.Y.N.

I tried to write a poem about turtles once.
No. Yes—slightly.
The turtles were soldiers, our helmets
Were our shells.
I don't remember the rest.
Maybe it was about how we marched,
In slow cadences towards death.
Rhyme, rhythm can make heroes and slaves of us all.
One way or the other my best friends didn't like it.
"Too obvious," once dismayed.
"I know you can do better," grumbled another.
But this remnant is the closest I can come to a thing of beauty.
Even the disaster of a poem has been wiped from my mind's eye.
Except.
For one image.
Hatchlings crackling out of their eggs, slowly marching towards
The sea. You see I still see them—the turtles. Yet, I and other soldiers
Were never them.
We were/are the birds of prey who peck at the sands.
The silent killers.
The deadly ones.

Nicole Goodwin, aka GOODW.Y.N,. is the author of the Warcries *(2016) and* Warcrimes *(2023), the latter from Atmosphere Press. The U.S. Army veteran served in Operation Iraqi Freedom. She is a 2018-2019 Franklin Furnace Fund Recipient, a 2018 Ragdale Alice Judson Hayes Fellowship Recipient, a 2017 EMERGENYC Hemispheric Institute Fellow, as well as the 2013-2014 Queer Art Mentorship Queer Art Literary Fellow.*

love sonnet to a new K-pot
by Randy Brown

You are a hard green bowl to crack apart,
inscrutable like Chinese egg-drop soup.
I trust to you my noodled self—not heart,
not groin—instead, my gray "brain-housing group."
In old steel pots, we troops could cook our grub,
or use the liners as a pail for brass.
We washed our socks and cocks in helmet tubs,
and settled on those tuffets head or ass.
Your greater weight now floats on donut foam,
and creases lines across my forehead bared—
with leathered sweatband held in place like Rome
once clipped a crown of thorns, my skull is snared.

But, fragile shell that's spun from Kevlar thread,
you have one purpose: Save my pounding head.

Randy Brown embedded with his former Iowa Army National Guard unit as a civilian journalist in Afghanistan, May-June 2011. A 20-year veteran with a previous overseas deployment, he subsequently authored the poetry collection Welcome to FOB Haiku: War Poems from Inside the Wire (2015). *He is the author or co-editor of multiple books and anthologies, including* Why We Write: Craft Essays on Writing War. *On social media, he often still writes about military themes, topics, and literature under his original mil-blogging pseudonym "Charlie Sherpa."*

P-38 Can Opener
by Benjamin B. White

I never took the time to verify the explanation,
but it was said the field tool was called a P-38
because of the number of turns it took to open
a can of cold ham or delicious pork and beans
coagulated in grease, and that tool was also
known as a John Wayne to help give soldiers
the heroic strength it took to eat a fruit cake
that had been sealed for the field since Korea
and left over after Vietnam to make sure Cold
War troops frozen in Germany could experience
the nutritional pain of deciding to live on cheese,
crackers, and peanut butter in packets that
needed to be kneaded to fold the natural oils
back into the gooey peanut putty part—but
then almost overnight, dehydrated MREs took
the place of the boxes of cans, and P-38s were no
longer necessary or issued even though they were
the most versatile tool the Army had—a can-
opener, sure, but also a screwdriver, a letter-opener,
a fingernail cleaner, a hole-punch, and overall,
a small, do-it-all toolbox in your pocket, and I
carried mine for years after I was discharged
from the Army and even when I subsequently
enlisted in the Coast Guard—the service I was in
when we joined the TSA—the Transportation
Security Administration—in the newly formed
Department of Homeland Security—DHS—just
a few years before I retired, so I always had a
connective feeling for that sister organization
and the TSA agents—until, that is, one day going
through an airport checkpoint, they questioned

my possession of a strange, little-over-an-inch-long,
metal object with a small, but sharp point, and a
dangerous looking notch meant to hook under
the lip of a tin can that I carried—had carried
for years on a key ring that once had been
a grenade pin—practical mementos and tools
of my service, but the agents all rushed over
to swarm around me to see what kind of terrorist
I was and how evil I must be to try to get that weapon
onboard a plane perhaps thinking I was going to use
a P-38 to open a cockpit door, which would have taken
a lot more turns than opening a can of tuna fish while
sitting in the German mud, but luckily, there was
an old soldier among their ranks who knew what
it was, understood why I carried it, explained
the value of it, and didn't let his young, inexperienced
coworkers confiscate it in a glorious act of saving
America from an attack of a veteran wielding a can opener,
so I got through the X-Ray machine, past the wand screening,
and beyond the gaze of their professional, stand-around-
all-day suspicion to make it onto the safety of the plane,
thinking about how that straight-off-the-street
generation of civilian government officials probably
didn't even know who John Wayne was.

Benjamin B. White was pursuing a career in baseball when a worn-out shoulder diverted his life's course. After an enlistment in the U.S. Army, White ended up as an officer in the U.S. Coast Guard. He served 22 years. Now retired, the poet and novelist lives near Albuquerque, New Mexico. He is the author of more than eight books, including Always Ready: Poems from a Life in the U. S. Coast Guard *(Middle West Press, 2022).*

Pop-Tarts, Rip Its & the Surge
by Chad Corrigan

My second deployment to Iraq was during the end of the Surge. If I was flying that day, I'd fill my pockets with Pop-Tarts and Rip Its on the way out of the DFAC. This is what would get me through my shift on the flight schedule. We had teams of two Apaches up 24/7. Show-time was 3 hours before wheels-up to pre-flight, get an ops and intel update, team brief and plan, eat quickly, and take off for 4 to 6 hours. I'd place my Pop-Tarts on the flat top part of the dash, with Rip Its tucked into the side where they wouldn't roll around. That mix of caffeine and sugar is how I sustained myself through those missions. You missed at least one meal every patrol.

I only remember eating strawberry Pop-Tarts. I'm not sure if we had other flavors. Rip It's came in three flavors: Power, Atom Pom (pomegranate), and sugar-free Citrus. Power was the O.G. and most common. I loved the first two and would grumble if they only had Citrus. Years later in Afghanistan, different FOBs had different flavors. You'd visit and find a new flavor, something you didn't know existed, like sugar-free grape.

Pop-Tarts and Rip Its went with me on every mission, just like my rifle and pistol.

While I grew to hate the DFAC, somehow, I never got tired of Pop-Tarts and Rip Its. I loved Rip Its—still do. However, I've only drunk a handful while stateside. Occasionally, I'll see one and buy it. But for me, it's a combat thing. A gas-station-bought Rip It is OK, but a free one pocketed from the DFAC, drunk when you're exhausted, is heaven.

The Surge was fought on Pop-Tarts and Rip Its.
At least it was for me.

Chad Corrigan is a U.S. Army officer and helicopter pilot. He holds an undergraduate degree in philosophy and political science from Stonehill College, Easton, Massachusetts. He also holds a graduate degree in public policy and management from the University of Pittsburgh. He is a former professor of military science and a department chair at Boston University, and a fellow at Harvard University. His writing has appeared in the Why We Write: Craft Essays on Writing War, *and in the journals* From the Green Notebook, The Wrath-Bearing Tree, *and* As You Were. *He has served in Iraq, Afghanistan, Syria, South Korea and Germany, and at the Pentagon. He is a member of the Military Writers Guild.*

Safety
by Amalie Flynn

I am still awake, in this new house, our bed, and my husband's arm, crossing over my chest, like a deadbolt.

And I think about the mechanism of a lock. The safety on the M4 my husband carried for one year in Afghanistan, locked but ready.

Or the way we sleep, *too often*, now, now that he is home, how we sleep, together, in our bed, but locked on opposite sides. Our hearts, this organ we assign too much to, or maybe, not enough, locked inside our rib cages.

Amalie Flynn is a poet and the author of September Eleventh: an epic poem, in fragments *(Middle West Press, 2021);* Wife and War: The Memoir *(2013); and a collection of poetry blogs:* Pattern of Consumption, September Eleventh, Wife and War, The Sustainability of Us, *and* Border of Heartbreak. *Her writing has appeared in* American Book Review, Beyond their Limits of Longing *(2022), the* New York Times, Time, *and* Huffington Post *and has received mentions from the* New York Times *and* CNN. *Flynn has an undergraduate degree in English and Studio Arts, a Master of Fine Arts in Creative Writing, and a doctorate in Humanities. She serves as poetry editor for* The Wrath-Bearing Tree *literary journal. Flynn lives in Rhode Island with her husband and their two children. She is currently working on a poetry collection focused on sexual and environmental violence.*

Thumper
by Charles Jacobson

The brass came by, proclaiming their message of war. "When the NVA pounces, we'll dump air and arty on him and wipe him out." Lt. Martinez, a veteran presence with four years in-country, didn't share the line. Martinez was of a mind that you had to be a little smarter than to raise a baiting operation in the Dog's Head. He spoke with amused vehemence as if he understood everything from the beginning. "Firebases are not a good place. No real cover, no room to maneuver, no chance to flank the other side."

I was wordless, wondering if the general staff will be in control, making smart decisions as the situation develops, though my immediate fear was an attack before Jay was hardened.

After the last-light patrol sallied forth, it was time to wrap. Col. Hannas, who was not above taking point, was there to check our night readiness like a good neighbor. He saw Thumper (my M79 grenade launcher):

"Are you ready to go?"

"Yes, sir."

"It's damn hot."

"Yes, sir."

"Do you need anything?"

"No, sir."

"Notice anything in the bush?"

"No, sir."

"I like your attitude. Show me what you can do." He pointed a hundred yards downwind. "See that fifteen-foot tree out there? Put a 79'er as close as you can."

The gangly sight on ol' Thumper was adios long ago (one less item to catch in the bush). Screw it. I gritted my teeth and used dead reckoning on a tree standing so erect at the edge of the wood that fluffing was out of the question. *PHOOT.* A direct hit ripped out a four-inch chunk,

sending bark and wood high in the air. *Nobody will fuck'n believe this.*
[monster applause]
"Well, you can't get any closer!"
Those in the woods were taking notes.

Charles Jacobson is a U.S. Army veteran, who has an abiding interest in philosophy and the arts and a cat who doesn't like him. The writer was a friend of Terry L. McClish, deceased, who was in Vietnam during 1969-1970, in a squad in Charlie Company, 2nd Battalion, 7th Cavalry, 1st Cavalry Division. The piece concerns the real-life combat experience of Sgt. McClish. Jacobson's work has appeared is such venues as: Proud to Be: Writing by American Warriors *anthology series,* Pure Slush Books, Fleas on the Dog, *Military Experience and the Arts'* As You Were *journal,* Poets Choice, Drunk Monkeys, Line of Advance, Wingless Dreamer, The Yard, *and Kallisto Gaia Press.*

Mr. Shingles
by Colin D. Halloran

They call me names.
Not like those kids in the back of the bus.
(Man, how time flies ...)
These are different:
Advent Calendar.
The Christmas Tree.
(TCT in an acronym-happy world)
And my favorite: they call me Mr. Shingles.
Pouches. Tan on grey.
One short of a dozen.
Strapped to the front of my vest.
No chocolates to be found inside.
0.23 kg—that's roughly half a pound—
H-E—that's High Explosive—
Rounds—grenades.
The joys of being grenadier.
Six extra pounds of volatility.
5-meter kill radius. 15m CR1.
These are the things you need to know.
Knowledge to be effective in the field.
Knowledge that one round to the chest
(a chest maybe 70% covered)
will take out you and anyone unfortunate enough
to be in that CR.
Boom.
So they call me Mr. Shingles.
And we laugh. I laugh.
Because what else can you do
when you are set to blow.

Note: The term "CR1" refers to "Casualty Radius." This is the area from the point of detonation, within which at least 50 percent of exposed personnel are projected to sustain injuries or death.

Colin D. Halloran is a U.S. Army veteran of Afghanistan. He authored the award-winning memoir-in-verse Shortly Thereafter, *as well as* Icarian Flux *and* American Etiquette. *He lives and teaches in Alabama.*

Improvised Explosive Device
by Liam Corley

> *"[Y]ou go to war with the army you have."*
> *—Donald Rumsfeld, U.S. Secretary of Defense*
> *1975-1977; 2001-2006*

The serpent line winding through the plank-floored
warehouse of the marsh-aired Fort Jackson distribution center
got me so pissed that I smacked my fist
into ceramic plate and hissed sotto voce,
"There's got to be a better fucking way."
Civilians every few yards supervised bins
overflowing with stained armor parts: chest plate, back
plate, groin shield, side panel, shoulder wrap, knee and elbow pads
in small, medium, large, extra-large, and a barely molded
female version to cause a little less pain in the breast.
Our parts came from at least a dozen soldiers
whose armor had returned from deployments before ours.
On this black-flagged South Carolina day, glistening drills barked,
    "Hydrate,"
throughout the battle rattle hour we spent
fumbling with molle straps and half-exhausted
Velcro fittings on Kevlar compartments. Now I understood
the bake sales for Dragon Skin Armor requested by buddies
who claimed that Army IBA failed after only one shot.

In Kabul, every broken piece of gear we kept
in makeshift service was marked TAYH, the army you have.
The Taliban mastered making do with scrap
we'd given the mujahideen and materiel Iran began to smuggle in.

When dismembered Humvees lolling off a smoking curb hit primetime,
SeaBee welders busted out some heavy plate and a Pentagon
logistics guy coined a magic phrase because it mostly worked:
"up-armored" trucks were all the rage no matter what the prefix said
about the canvas doors we had at first.

I sit now with a writer who is blocked, and I think
Rumsfeld could have been our secretary of poetic
state. The Taliban are still kicking ass; even their ghazals
explode. Take up your pack, friend. All our MFAs
went to Iraq, and the rotator for Kandahar leaves tonight. Go to war
with the words you have. Magazines will come after.

Liam Corley is a currently serving U.S. Navy reservist with deployment experiences in Iraq, Afghanistan, and throughout the Pacific area of operations. In his civilian career, Corley is a humanities professor at California State Polytechnic University, Pomona, Calif. His debut poetry collection, Unwound: Poems from Enduring War, *was published by Middle West Press in 2023.*

bullet proof me
by Randy Brown

"with your shield or on it,"
Spartan women told their men.

we bought my body armor online
from a site called "bullet proof me" dot com.
cost my wife and me a couple of grand
when I left to write the war.

be sure to read for guarantees, I joked.
the carapace unpacked as pieces
sheathed in bubble wrap.

the chestnut vest is stiff with heavy plates
that might stop a few machine gun rounds,
but will also break if you drop it wrong.

mine did its job and kept me safe.
I brought it back uncracked, and good as new
to the land of no refunds and no returns.

it's not like granddad's musket—
a thing to place over one's mantel
to inspire memorable conversation.

instead, it's curled up in a box
the dry husk of an un-embedded bug.
an aegis of change—all sales were final—
a shell of my immortal self.

Randy Brown embedded with his former Iowa Army National Guard unit as a civilian journalist in Afghanistan, May-June 2011. A 20-year veteran with a previous overseas deployment, he subsequently authored the poetry collection Welcome to FOB Haiku: War Poems from Inside the Wire (2015). *He is the author or co-editor of multiple books and anthologies, including* Why We Write: Craft Essays on Writing War. *On social media, he often still writes about military themes, topics, and literature under his original mil-blogging pseudonym "Charlie Sherpa."*

I wake to drowning
by Ben Weakley

in air. I sweat my salt
to the midnight moon.

Each night I put my head down
and cover this body
in high thread count armor.

I fight God
and the Devil in my dreams
until—waist-deep

in dog-tags and empty brass—
I run out of friends and ammunition.

Ben Weakley is a U.S. Army veteran of Iraq and Afghanistan, as well as a desk in the Pentagon. His debut poetry collection, HEAT + PRESSURE: Poems from War, *was published in 2022 by Middle West Press. He is a past poetry winner in the Col. Darron L. Wright Memorial Awards, as well as the Heroes Voices' Awards. He lives in Tennessee.*

In the Back of a Deuce-and-a-Half
by Bill McCloud

We were on the way to
the firing range on a
sweltering Fort Polk afternoon
in the back of deuce-and-a-halfs
Everybody in our truck
was laughing and joking
and my buddy was just
messing with the magazine
to his M14 rifle

All of a sudden a spring
came flying out of it to
hit him smack in the mouth
You could tell it hurt
hurt bad cut his lip
and while we watched
with him doing everything he
possibly could to prevent it

he started to cry
He obviously hated the fact
that he couldn't stop crying
The rest of us immediately
sobered up and were totally
quiet as the trucks began
pulling up to the range

Bill McCloud is a U.S. Army veteran. He is the author of two books on the Vietnam War: A collection of poetry, The Smell of the Light, *and an oral history titled* What Should We Tell Our Children About Vietnam?

Hot Landing Zone
by Dennis Maulsby

> *"Troops employed in the securing an LZ [Landing Zone]*
> *are highly vulnerable to VC/NVA attack, especially when the first*
> *helicopter troop lift is small because of a restricted LZ."*
> *—handbook for U.S. Forces in Vietnam*

thrumthrumthrum
rushing blurred-canopy jungle

dripdripdrip
stinking sweat-oiled bodies

chukchukchuk
hovering flesh-cutting guns

swikswikswik
crossing spin-burning tracers

whumpwhumpwhump
exploding skin-flaying shrapnel

mommommom
screaming bloody-gutted friend

movemovemove
son of a bitch!

Dennis Maulsby is a retired bank president and U.S. Army veteran living in Ames, Iowa. He is the author of an award-winning book of war poetry, Near Death/Near Life *(2015), a book of short stories,* Free Fire Zone *(2016), as well as others. Visit: www.dennismaulsby.com*

Blanket Party
by Martin Ott

In Fort Leonard Wood, our rooms were windowless,
the days begun in the dark for push-ups on fields of rock.
We were calorie-starved with only minutes to shovel
chow, and set against other squads by barking sergeants.
Allergic to wool and blistered by the sun, my skin
was tattooed red, the only color that seemed to matter.
Private Hawkins was the squad loudmouth and ox,
who bullied his way to the top bunk by the showers.
We may have been newbies in the ways of dying,
but understood the strange power we would wield.
When Hawkins emerged the next night from the john,
a blanket landed on his head, followed by darkness
as lights were cut. Grunts from kicks and punches
pulled us all out to view the writhing lump of man.
Wordlessly, we were all given the choice to take
one blow, and this was our first test of cowardice.

*Martin Ott is the author of ten books of poetry and fiction, the former
including* Captive *(C&R Press, 2011);* Underdays *(University of Notre
Dame Press, 2015); and* Lessons in Camouflage *(C&R Press, 2018) A
former U.S. Army interrogator and longtime resident of Southern
California, Ott works as a communications professional and develops for
TV and film in-between other writing projects. His latest novel,* Shadow
Dance, *is a noirish narrative about a U.S. veteran who gets caught up in
the dark underworld of an Iranian-run gentlemen's club in Los Angeles.*

Ode to a Pineapple Grenade
by Gerardo "Tony" Mena

Your fingers as lightning bolts
spindle across the sky,
cull skin into a pile,
a rake combing dirt,
softly singing,
All along the watchtower
princes kept their view.
Resurrect your Apocalypse,
sleeping in her white shroud
and nestled in mineral
earth, because war is a performance,
and then we bow.
Black. Curtains. Fire.

Gerardo "Tony" Mena is an Operation Iraqi Freedom veteran who spent six years in the Special Operations community with the Reconnaissance Marines as a Special Amphibious Reconnaissance Corpsman (SARC). During his tour in Iraq, he was awarded a Navy Achievement Medal with a "V" device for multiple acts of bravery. His debut poetry collection, The Shape of Our Faces No Longer Matters, *was published by Southeast Missouri State University Press in 2014.*

As I Clean My Rifle
by J.B. Stevens

I'm still tired from that last patrol
The quiet holds me,
And the flow state hits.
I could do this in my sleep—
Zen through rifle disassembly,
A blindfold party trick that no one cares about.
The warm embrace of CLP,

Cleaning, lubricating, protecting—a wondrous product—
The smell knocks and I spit out the taste of
Sand, and IED aftershocks rattle in my chest.

I loved my M4,
And I love my AR15. Mine.
Ecstasy through reassembly.

The dry heat comes and the never-
Ending exhaustion. I'm still tired from that last patrol;
I always will be—a problem that no one cares about.
The last patrol. An existence no one cares about.

J.B. Stevens is a writer of fiction, non-fiction, and poetry, often delivering hardboiled action, gritty conflict, and dark humor. His collection of noirish short-fiction, A Therapeutic Death: Violent Short Stories, *was published in 2022 by Shotgun Honey Books. Infused with TV and other pop-culture references, Stevens' debut poetry collection,* The Best of America Cannot Be Seen: Pop Poems, *was published in 2021 by Alien Buddha Press. Recently,* The Explosion Takes Both Legs: Noir Poems from the War in Iraq, *was published by Middle West Press. Stevens is a two-time past finalist in the Col. Darron L. Wright Memorial Writing Awards.*

CBR
by Dale Ritterbusch

for the M17A1 Protective Mask

At the end of the day,
after all that good training—
first the block of solid
instruction in the bleachers,
the practical demonstrations,
all that practice with the mask:
putting it on, clearing it, testing the seal,
looking like alien creatures
from some Grade B movie in the Fifties,
and then the gas chambers—
first CS, going in masked,
black rubber tight to the face,
flutter valve sucking in and out
with each labored breath, and always
the smell, the taste, of rubber and plastic
swimming in that swirl of smoke,
dark, highly contaminated,
an atmosphere from some other planet
hot and uninhabitable,
then taking it off, reciting for the Sarge
name, rank, serial number—
maybe a question from the twelve standing orders—
and then, with tears streaming down
they're let loose into the fresh, warm
air that clears their eyes and lungs,
brings home the truth that the mask works.

Next, after a short break, the second chamber,
chlorine gas this time,
just like World War I, and they

go in unmasked, double time,
line up inside, against the walls,
two squads at a time, adrenaline pumping,
waiting for the command to mask,
the admonition to clear it
before taking a breath—and one
kid, trying to cheat, had unbuttoned
his canvas bag, and his mask had fallen out,
chlorine turning his brass a sickly green
like his face, his panicking eyes
wild with fright. I looked at him, made him
stand there for a moment and think about it.

Afterwards, when the young recruits
thought the training was over, and they're
standing in formation, given the order
to smoke 'em if they've got 'em,
we'd see what they had learned,
popped four CS grenades
clamped at the ends of mop handles
and ran around the troops, surrounding
the formation with gas,
and some of them stood, put on
their masks as they'd been trained to do,
cleared them, ready for anything
while others ran, dropping equipment
everywhere, rifles, canteens, helmets, pistol belts
and more, and the drill sergeants
were pissed—had to round up
their troops, making them late for dinner.
And this one guy climbed up a tree,
poured water from his canteen
over his head and called for his momma:
and the sergeant yelled, "Get your white

ass out of that tree you dumbfuck
or I'll show you your momma."
And then they'd take all that equipment
dumped on the ground and throw it
in the CS chamber, in a pile, and make all
those dumbfucks go in there and sort it out,
making sure they picked up their own
rifles, and those sergeants would check
the numbers to make sure.

And this one kid just sat there
in the dirt, kneeling, coughing, gas still
thick in the air, and I walked over to him
and held that smoking grenade under his face
and yelled, "Put your mask on," and
he looked up at me, tears running out
of his bloodshot eyes and says, "I've
only got one lung," and I say, "Then
you'd better save that one goddam lung
and put that mask on," and I held
that grenade there, not believing anyone
could be so dumb not to use what he
has learned, knowing that mask would save
all that pain, and yet he wouldn't
put it on and I wouldn't put down that grenade—
And in my dreams I still stand
there, grenade smoking on the end of that mop handle
holding it close to his face, making him
learn and learn and learn and I'd
still do that today, knowing no matter
how good the training, we never learn anything,
that ignorance brings its own reward,
that I'm still standing there.

Dale Ritterbusch served in the U.S. Army from April 1966 to September 1969, as both enlisted soldier and officer in the Chemical Corps. As a liaison officer attached to JUSMAAG/MACTHAI in 1969, he was responsible for coordinating classified shipments of anti-personnel mines used along the Ho Chi Minh Trail and other infiltration routes.

The author of four collections of poetry, he recently retired as a professor of English at the University of Wisconsin-Whitewater. He was twice selected to be the distinguished visiting professor in the Department of English and Fine Arts at the United States Air Force Academy, Colorado Springs, Colorado. His creative work is archived in the Department of Special Collections at La Salle University.

Life and The First Gulf War
by Juan Manuel Pérez

Sonnet No. 15

Thirty years ago, serious training
the worst thing that could happen in a war
from nuclear to biological to chemical; setting up for the worst
that man could muster against fellow man
that sophisticated stupidity
a congress throwing about monkey-shit
recusing themselves from their former selves
distancing intelligence by long miles
so here we are now, ten months since removed
the longest Spring Break nobody wanted
behind computers like futures of old
wrapped in shreds of bureaucratic red tape
… give me back my MOPP gear, I want to live

Juan Manuel Pérez, a Mexican-American poet of Indigenous descent and the Poet Laureate for Corpus Christi, Texas (2019-2020), is the author of numerous poetry books including Thirty Years Ago: Life and the First Gulf War *(The House of the Fighting Chupacabras Press, 2023). Juan, a former migrant worker, is also the 2021 Horror Authors Guild's Inaugural Lifetime Achievement Award winner and a recipient of a 2021 Horror Writers Association Diversity Grant. To learn more about this award-winning poet, combat veteran, teacher, and gourd dancer, visit:* www.juanmperez.com

The Tranquil Fires in the Boat
by GOODW.Y.N.

The fireworks explode like distant mortar rounds.
The light bouncing high into the air never descends to earth again. It is
    only I trapped w/ this cage,
Carved out of my own flesh and silent tears.
As the audience collects along the shore
dumbfounded; the flickers began to flourish within the boat.
As the emptiest seashells scattered decorating the beach.
My mind is buried in the sands a million years past and away.
Where the celebrations were not so friendly for, they cherished war and
    nurtured death
And the people's eyes held what was embodied in them, shock   awe
    fear
For we had hard-bludgeoned it into their heads, similar to how we plug
    the pledge of allegiance fresh into a child's mind
Soldier and death had become one in the same
As for the rest of America, it celebrates its liberation.
I am not of the stock that feels proud.
I am not of the trade that feels liberated.
I am the shadow behind the American flag, when it is half-masted and
    mostly forgotten
Every other day of the week

Nicole Goodwin, aka GOODW.Y.N,. is the author of the Warcries *(2016)
and* Warcrimes *(2023), the latter from Atmosphere Press. The U.S. Army
veteran served in Operation Iraqi Freedom. She is a 2018-2019 Franklin
Furnace Fund Recipient, a 2018 Ragdale Alice Judson Hayes Fellowship
Recipient, a 2017 EMERGENYC Hemispheric Institute Fellow, as well as
the 2013-2014 Queer Art Mentorship Queer Art Literary Fellow.*

Pen Flares
by J.B. Stevens

Pen Flares are crimson streaks of happiness,
  in a colorless world.

One-hitter Roman candles.

They are given to young private Soldiers as a deterrent,
Shoot the streak at an approaching bad guy,
A warning (so you don't have to shoot him in the face with
  a 5.56 round).
They are good for that,
But young private Soldiers use them for fire-works fights, Shooting one
  another,
Or shooting camel-spiders,
Or screwing around.
Pen flares are joyful,
And I saw numerous young men almost light one-another on fire while
  laughing and playing in a warzone.
Laughter in a war zone is a precious thing (even if you might light your
  battle buddy on fire).

With little red happy veins.

I often think of them and smile.

J.B. Stevens is the author of a collection of hardboiled short-fiction, A Therapeutic Death: Violent Short Stories, *published in 2022 by Shotgun Honey. His debut poetry collection,* The Best of America Cannot Be Seen: Pop Poems, *was published in 2021 by Alien Buddha Press. Another poetry collection,* The Explosion Takes Both Legs: Noir Poems from the War in Iraq, *was recently published by Middle West Press.*

The Sea-Farer Flies
by Jan Gunter

Authority is meant to be denied just as whims are meant to be satisfied. That plane wasn't his, but the sleek navy blue reflecting the mid-day sun inspired a desire that was *all* his.

Frank was going to take her for a ride.

How would he do it? With God watching over, of course, because he hadn't a clue how to fly it.

When an opportunity alone struck, so did he; he struck as the cobra. That plane was his lunch, for when he landed it—if he landed it—his sergeant would see to denying him dinner in the mess hall.

The cockpit smelled of fresh leather and heavy-duty uniforms. That dense scent had become as common as his own sweat in recent months. Hints of water and salt drifted off the ocean, its winds pushing him inside and urging him to don the headpiece.

Enthrallment carried him across the borders that his knowledge met. All boys his age knew a thing or two about planes, but looking at a toy in the Sears magazine before Christmas was a little different than this.

Frank felt the same nervous excitement he had when he drove his father's car for the first time. He was 13 and all legs in the driver's seat. If Pa could see him now, he would be very disappointed. The feeling overwhelmed him with joy.

Planes were a lot like boats, he found, except boats didn't crash quite as hard as that plane did. His muscles braced for impact, the sides squealed as he landed it. Metal scraped the unfortunate building he had just rubbed up against, a cat looking for fond words from an owner.

His jelly legs propelled out of the cockpit; all he earned was trouble.

Jan Gunter is a student and freelancer based in Ohio. Frank was Gunter's beloved grandfather, a U.S. Navy veteran who fought in Korea, and whose shenanigans inspired a lifetime of creativity.

Slipping the Surlies
by Eric Chandler

after "High Flight" by John Gillespie Magee, Jr.

Oh! I've slipped the reflective belt and dirt,
And danced the skies on my dust-covered wings;
Sunward I've climbed, and tried to stay alert
With my go pills and flew a thousand rings
You have not dreamed of—wheeled over the dung
High in light-brown violence. Capping there
I've chased the stinking wind along, and flung
My eager craft through sandstorms in the air ...

Up, up the long, dry, afterburning view
I've topped the piddle-pack with easy grace.
Which drinking a Rip-It will make you do—
And, while with silent, drifting mind I've trod
In a narrow altitude block of space,
—Put out my hand, and slewed the Sniper pod.

Eric Chandler is the author of the poetry collections Kekekabic *(Finishing Line Press, 2022) and* Hugging This Rock: Poems of Earth & Sky, Love & War *(Middle West Press, 2017). He is a three-time poetry winner of the Col. Darron L. Wright Memorial Award, and a retired veteran of the U.S. Air Force and the Minnesota Air National Guard. Over Iraq and Afghanistan, he flew 145 combat missions in the F-16. He is happiest when on a Duluth, Minnesota trail with his wife, two children, and dog Leo.*

Snack Time's "Clicks and Snaps"
by Stan Gaidis

In 1970, I served in the U.S. Air Force and deployed to Vietnam as part of a radar detachment.

Before going to bed, I regularly "geared up" in my flak vest and helmet, loaded my M16, and pocketed four spare mags. Then I walked across base to the Aerial Port (Det 5, 8th Aerial Port Squadron), where the small snack bar sold fresh(er) hot dogs, hamburgers, and other sundries for early-departing passengers.

At the doors to the terminal, I made my weapon "safe" by removing the magazine from the rifle, racking the slide back, letting it snap forward, inserting the muzzle through the 3-inch hole in the lid of the red sand-filled barrel (set at a 45-degree angle), and pulling the trigger. The "click" meant the weapon was rendered "safe" from accidental discharge.

At the snack bar, I selected four hot dogs from the steam table and doctored them up with mustard and chopped onions. I paid the attendant with the Military Payment Certificate—quasi-currency that we called "funny money"—then reloaded my M16 and walked back across base in the humid jungle night.

Once at my quarters, I put my gear away, snagged a beer from the reefer in the day-room, and sat down to enjoy my "down home" snack.

My tour in Vietnam eventually ended. I transitioned home and transitioned out; I married my wife and we raised our children. What once reminded me of home eventually reminded me of war. I can still hear the clicks and snaps.

Even today, my family sometimes asks why I enjoy hot dogs steamed and doctored a specific way.

"Because I like hot dogs," I always say. I never tell them about the clicks and snaps.

Until now.

Stan Gaidis is retired after a long U.S. Air Force career that included serving in the Strategic Air Command during the Vietnam War, creating a device that traveled on the Voyager II probe out of our solar system, driving a semi-truck, and working in information technology. He lives and writes in Indiana, with his wife, Janet.

The Acme No. 470 Clicker
by Jehanne Dubrow

after the signaling device carried by paratroopers
in the US. 101st Airborne Division on D-Day, June 6, 1944

The tool was pocket-sized, designed to rest
between a soldier's fingers and his thumb.
A toy. But when the metal was compressed
against the brass, it made the voice of some
lost insect in the dark. He clicked it once
to ask, are you an enemy or friend,
so brief the sound was barely utterance.
He listened for a clicking back. To spend
the night crouched in uncertainty, the rush
across the shadows, hours of pushing past
long grass—it's hard imagining the hush,
and then the stridulation, low and fast,
that, for a moment, quieted his doubt,
the little noise of crickets chirping out.

Jehanne Dubrow is the author of nine poetry collections, including most recently Wild Kingdom *(Louisiana State University Press, 2021), and two books of creative nonfiction,* throughsmoke: an essay in notes *(New Rivers Press, 2019) and* Taste: A Book of Small Bites *(Columbia University Press, 2022). Her third book of nonfiction,* Exhibitions: Essays On Art & Atrocity, *was published by University of New Mexico Press in 2023. Her writing has appeared in* POETRY, New England Review, Colorado Review, *and* The Southern Review. *She is a professor of creative writing at the University of North Texas.*

The Last Four
by Lucas Randolph

My son hates school. When he started carrying my old military ear defenders around, the hard plastic kind they first issue you when working on the flight line, it seemed to make things easier. Double sided foam and jet-black plastic—rated for 29 decibels, at least. He carried them everywhere, a symbol of pride. The school wanted to know where they came from, and I lied saying they were an old work pair—not really a lie, I suppose.

They said my son has sensory issues. I could at least relate to some aspect of that. I didn't care much for loud noises anymore, either. Couldn't stand them, in fact. Didn't really seem to have a good reason for it, though—not like him, anyway.

When my son asked if there was something wrong with him—I said no, he was perfect just the way he was.

A therapist from school gave him one of the squirrely looking headsets, you know the type—colors that scream bright and a little cartoon animal with a conversation bubble, so he can write his own name on it. Instead, he wears the last four of his dad's social security numbers around his neck, etched into the bottom with an engraver for accountability reasons.

Accountability. Accountability is important. They tried teaching me that while I was in the service. I would try to teach that same lesson to my son:

Don't lose those—that's our only pair.

I get mad whenever he does.

Ear defenders were an aircraft mechanics most important tool on the flight line—a lost tool could ground a jet—or worse. Now, I'm constantly terrified my son might lose them—or worse.

No God Damn accountability.

I've started to suspect I might be the one with something wrong.

Lucas Randolph is a veteran of the U.S. Air Force and served as an avionics technician from 2005-2014. He holds an undergraduate in English and creative writing and a Master of Fine Arts in fiction from Antioch University Los Angeles. His literary writing has previously appeared in such venues as As You Were, The Wrath-Bearing Tree, *and* Unlikely Stories Mark V.

Plastic
by Aramis Calderon

I dreamt of America on the flight line.
The engines idled under the Arabian sun.
We cooked inside our ride home.

Our armored bodies packed close.
Rifle muzzles up.
Bitching.
Cursing.
Complaining.

Our time had come to go.
Someone sang
"C-130 rolling down the strip ..."
We gained altitude
above the desert.

No one spoke further.
Without cabin pressure
everyone sounded faraway.

My plastic spit bottle
Collapsed.
Then I dreamt of Iraq.

Aramis Calderon is a U.S. Marine Corps veteran and the author of the novel Dismount *(A15 Publishing, 2019), and the forthcoming memoir* Fugitive Son *(Potomac Books, 2024). He holds a Master of Fine Arts in Creative Writing from the University of Tampa. His current area of operations is Tampa Bay, Florida, where every week he meets with fellow veteran writers in the DD-214 Writers' Workshop.*

Acknowledgements

We are grateful to the editors and publishers of the following magazines, journals, and collections in which these poems first appeared—some in slightly different versions. (Poems listed alphabetically by Author's last name.)

David Abrams' "Great Valentine's Day Barrier Riff" originally appeared as an untitled post on the author's Facebook page, Feb. 14, 2023.

Aly Allen's "Recall" first appeared in the chapbook *Approaching Valhalla* (Bottlecap Press, 2022).

Jessi M. Atherton's "Guacamole," "Last Four," and "Pink Elephant" first appeared in *The Time War Takes* (Middle West Press, 2023).

Callie S. Blackstone's "Your Future Love" first appeared in *Freshwater Literary Journal*, 2021.

Randy Brown's "bullet proof me," "dust bunnies and combat boots," "love sonnet to a new K-pot," and "your squad leader writes haiku" first appeared in *Welcome to FOB Haiku: War Poems from inside the Wire* (Middle West Press, 2015).

Eric Chandler's poem "Man-Machine" appeared in *Hugging This Rock: Poems of Earth & Sky, Love & War* (Middle West Press, 2017). His "Slipping the Surlies" first appeared in *Proud to Be: Writing by American Warriors, Vol. 6* (Southeast Missouri State University Press, 2017).

Liam Corley's "Improvised Explosive Device" in *The Line Literary Review*, Spring 2021. "Something Else You Don't Need" first appeared in *Proud to Be: Writing by American Warriors, Vol. 1*, (Southeast Missouri State University Press, 2012).

Jehanne Dubrow's "The Acme No. 470 Clicker" was first published in *The Hudson Review*, Summer 2022. "Trench Whistle" was first published in *Los Angeles Review*, Oct. 3, 2022. "Non-Essential Equipment" was first published in *Stateside* (Northwestern University Press, 2010).

Amalie Flynn's "Safety," "Silver Dollar," and "Boots" are excerpts previously published in *Wife and War: The Memoir* (2013).

Aaron Graham's "The Boots on the Ground" first appeared in *Blood Stripes* (Sundress Publications, 2019).

Nicole Goodwin's (a.k.a. GOODW.Y.N.) poems "The Tranquil Fires in the Boat" and "The Unholy Cabal" first appeared in *Warcrimes* (Atmosphere Press, 2022).

Colin D. Halloran's "Bottle Opener, or 'Sestina for my CIB,'" "Footlocker," and "Mr. Shingles" were first published in *Shortly Thereafter* (Mint Hill Press, 2012).

Farzana Marie's "Meeting in Amman" first appeared in *As You Were*, the literary journal of Military Experience & the Arts. Vol. 3, Fall 2015.

Dennis Maulsby's "Enlightenment" and "Hot Landing Zone" were first published in Near Death / Near Life (Prolific Press, 2016).

Gerardo "Tony" Mena's "Holding Maps" and "Ode to a Pineapple Grenade" first appeared in *The Shape of Our Faces No Longer Matters* (Southeast Missouri State University Press, 2014).

Peter Molin's essay "The Leatherman" first appeared in *The Wrath-Bearing Tree* May 1, 2023.

Abby E. Murray's "Army Ball" and "Jewelry" first appeared in *Hail & Farewell* (Perugia Press, 2019).

Martin Ott's "Battlefield Typewriter" and "Blanket Party" first appeared in *Underdays* (University of Notre Dame Press, 2015).

Juan Manuel Pérez's sonnets Nos. 15, 43, and 45 first appeared in *Thirty Years Ago: Life and The First Gulf War* (The House of the Fighting Chupacabras Press, 2023).

Susanne Rancourt's "who dares, wins" first appeared in *Songs of Archilochus* (Unsolicited Press, 2023).

Dale Ritterbusch's "CBR" first appeared in *Lessons Learned* (Viet Nam Generation Inc. and Burning Cities Press, 1995).

Mason Rodrigue's "Black Bracelets" and "EGA" first appeared in *Rock Eater* (Dead Reckoning Collective, 2023).

Karen Skolfield's "Battle Dress Uniform" was first published in *Poetry Magazine*. "Boots: *Origin < Old English*, remedy, fortunate" was first published in *The Plume 7 Anthology of Poetry*.

J.B. Stevens' "As I Clean My Rifle" and "Pen Flares" first appeared in the chapbook *All the Violent Memories* (Close to the Bone Publishing, 2021).

Lisa Stice's "Wedding Arch" and "While in Uniform" first appeared in *Uniform* (Kelsey Books / Aldrich Press, 2016).

Ben Weakley's "I wake to drowning" and "The Wooden Elephants of Herat" first appeared in *HEAT + PRESSURE: Poems from War* (Middle West Press, 2022). "Soldier's Song" first appeared in the literary journal *Line of Advance* journal, it won a second-place in poetry in the 2020 Col. Darron L. Wright Memorial Writing Awards.

Benjamin B. White's "Dress Pants," "Gig Line," "Sextant," and "Uniform Gedunk" first appeared in *Always Ready: Poems from a Life in the U.S. Coast Guard* (Middle West Press, 2022).

Glossary

AK-47: A "Kalashnikov" is any one of a series of rifle models based on original designs by Russian arms inventor Mikhail Kalashnikov (1919-2013). The "Avtomat Kalashnikova" (A.K.)—"Kalashnikov's automatic rifle"—was introduced to the Soviet military in 1947. It fires a 7.62mm round.

AR15: The civilian semi-automatic variant of the M16 assault rifle. The abbreviation "A.R." stands for "ArmaLite Rifle," after the original manufacturer of the weapon.

A.P.C.: Armored Personnel Carrier

B-4 bag: A military-issue or -styled garment bag; a fold-over piece of luggage capable of carrying full military dress uniform, as well as 2 to 3 pairs of civilian clothes, shaving kit, shoes, boots, and undergarments.

C-123 "Provider": The C-123 "Provider" is a two-engine propellered military cargo aircraft manufactured by Fairchild Aircraft, and first built in 1956. While serving many roles, it is often remembered for its use in aerial spraying of defoliants and insecticides.

C-130 : The C-130 "Hercules" is a four-engine turboprop military cargo aircraft manufactured by Lockheed Martin, first built in 1954. One C-130 can carry up to 92 passengers, 72 combat troops, or 64 paratroopers.

C.B.R.: "Chemical, Biological, Radiological"

C.L.P. ("Cleaner, Lubricant, Preservative"): A type of a liquid issued to U.S. military personnel for maintaining small arms, such as rifles and pistols. Sometimes called "Break-Free," which is a registered trademark.

C.S.: CS "gas" is a chemical, non-lethal riot-control agent ("tear gas") created by aerosolizing the chemical compound of 2-chlorobenzalmalononitrile. First synthesized in 1928 by Americans Ben Corson and Roger Stoughton, the product's name is derived from the first letters of each creator's surname.

C.V.N.: The 3-letter U.S. Navy hull classification for "Carrier, Volpane, Nuclear"—a nuclear-powered aircraft carrier.

CR1: "Casualty-Inducing Radius." The radial distance at which a grenade may cause death or injury. The M67 fragmentation grenade has an effective "kill zone" radius of 5 meters (16 ft.), while the casualty-inducing radius is approximately 15 meters (49 ft.).

DFAC ("DEE-fak"): A military dining facility. In some eras and branches, also called a "mess hall" or "chow hall."

D.I.Y.: "Do It Yourself"

D.M.Z. ("Dee-Em-Zee"): De-Militarized Zone. Any area that cannot be used for military purposes, by treaty or other agreement. A generic term, but often used in reference to space between South Korea and North Korea, established by an incomplete 1953 armistice agreement.

D.S.: "Drill Sergeant"

"Deuce-and-a-Half": Nickname for a 6-wheeled M35 series 2-and-a-half-ton cargo truck.

E.G.A.: "Eagle, Globe & Anchor." The emblem of the U.S. Marine Corps.

H.D.: "High-Definition"

H.E.: "High-Explosive"

HESCO ("HESS-koh"): A manufacturer of modular bag-and-cage barrier systems, which, when filled with sand or other material, can be used to protect military facilities.

I.E.D.: Improvised Explosive Device. A "homemade" bomb or mine constructed of parts, including military scrap or surplus, not originally intended for such use.

JUSMAAG: "Joint United States Military Advisory and Assistance Group"

L.Z.: "Landing Zone"

M14: The wooden-stocked M14 ("em-fourteen") was the standard semi-automatic battle rifle issued to U.S. soldiers between 1957-1967.

M4: The M16 ("em-sixteen") family of semi-automatic assault rifles has been issued to U.S. Army soldiers since 1969. It fires a 5.56mm round, usually as either single shots or three-round "bursts." A shorter version, the M4 carbine, gradually replaced the M16 rifle starting in 1994.

M79: The M79 ("em-seventy-nine") was a single-shot, breech-loading, shoulder-fired 40mm grenade-launcher used by U.S. forces during the Vietnam War. It was known by various nicknames, including "Thumper" and "Blooper," the latter because of the distinctive sound made when it was fired.

MACTHAI: Abbreviation for "Military Assistance Group in Thailand."

Mameluke sword: A slightly curved, scimitar-like sword traditionally worn by U.S. Marine Corps officers in dress uniform.

M.F.A.: "Master of Fine Arts." A type of graduate-level academic degree, often focused on creative writing.

M.I.A.: "Missing In Action"

MOLLE ("MAW-lee"): "MOdular Lightweight Load-carrying Equipment." A standardized systems of packs, pouches, straps, and attachments used by the U.S. Army and other military organizations.

M.O.S.: "Military Occupational Specialty." An alphanumeric designation or job code identifying a military service member's training and job fun

MOPP ("Mawp"): "Mission-Oriented Protective Posture." Military equivalent of Personal Protective Equipment (P.P.E.) intended for use in tactical environments.

NATO ("Nay-toh"): "North Atlantic Treaty Organization"

N.C.O.: "Non-Commissioned Officer"

N.V.A.: North Vietnamese Army

O.E.F.: Operation Enduring Freedom

O.G.: "Original Gangster"

O.I.F.: Operation Iraqi Freedom

P-38: The P-38 is a pocket-sized (1.5 inches long), metal can-opener first issued to U.S. military personnel in 1942. It consists of a short handle, with a small, hinged metal tooth that folds out to pierce the can lid. It is often called a "John Wayne," after the rugged, 20th century American movie star of the same name.

P.C.S.: "Permanent Change of Station"

P.T.S.D.: Post-Traumatic Stress Disorder. A medical condition stemming from exposures to actual or threatened death, serious injury, or sexual violence—including exposures through the experiences of others—that result in persistent negative alterations of an individual's thinking, emotions, sleep, and other behaviors. More-formally described in the American Psychological Association's *Diagnostic and Statistical Manual of Mental Disorders (DSM).*

P.X.: "Post Exchange." An exchange is a retail store on a military installation that sells to military personnel and authorized civilians.

Rip-It: A brand of energy drink produced by National Beverage Corp., Fort Lauderdale, Florida.

SFAT ("ESS-fat"): "Security Force Assistance Team." A 12-member cadre of U.S. Army soldiers assigned to mentor, enable, train, and accompany military units of partner nations. SFAT were first formally established in 2017; ad hoc units called Military Transition Teams ("MiTT") and Embedded Training Teams (E.T.T.) fulfilled similar missions during Operations Iraqi Freedom and Enduring Freedom-Afghanistan, respectively.

S.A.R.: "Search-and-Rescue"

T.A.Y.H.: "The Army You Have." During a December 2014 public Q&A session while visiting with U.S. troops in Kuwait, U.S. Secretary of Defense Donald A. Rumsfeld responded to a question from U.S. Army Spc. Thomas "Jerry" Wilson. Wilson had asked about why he and his fellow soldiers were scrounging for scrap metal in order to increase their unarmored wheeled vehicles' resistance to bullets and shrapnel. A widely reported version of Rumsfeld's response was interpreted by many as brusque: "You go to war with the army you have, not the army you might want or wish to have at a later time."

T.D.Y.: Military orders code for "Temporary Duty"

V-42: A double-edged fighting knife first issued to a joint U.S. and Canadian commando unit in World War II.

V.A.: The U.S. Department of Veterans Affairs

Discussion & Writing Prompts

Topic: "Taking Flights"

In "Slipping the Surlies" (page 155), former U.S. Air Force F-16 fighter pilot Eric Chandler cheekily echoes "High Flight," a famous 1941 poem by Spitfire pilot John Gillispie Magee Jr. The latter poem ends with the poet touching "the face of God." In "The Sea-Farer Flies" (page 154), Jan Gunter tells a story about their grandfather Frank, a sailor who legendarily borrowed a U.S. Navy plane for a short joyride. In "Hot Landing Zone" (page 143), Vietnam War veteran Dennis Maulsby describes a military airlift in visceral, sensory details: sounds, smells, and chaos. In "Snapshot" (page 92), poet Layle Keane Chambers captures the weight of a parent's fears on the day of a young U.S. Air Force pilot's first solo flight. In Juan Manuel Pérez's "Life and the First Gulf War, Sonnet No. 45" (page 125), the beauty of blue sky viewed from a helicopter sparks an uncomfortable childhood memory.

Question No. 1:
In each of these examples (or others), what sensory descriptions are notable to you? Which do you find most-resonant, and why?

Question No. 2:
What does the experience of flight represent or reveal to each of these writers? Escape? Diversion? Possibility? Responsibility?

Writing Prompt:
Write about a time you felt untethered or transported, whether physically or emotionally or both. Alternatively, write about a time you felt "grounded." For example, perhaps this was a time when you felt full of confidence (or full of dread), or particularly rooted to a place or spot.

Topic: "Breaking the Codes"

In "who dares, wins" (page 59), U.S. Army and Marine veteran Suzanne S. Rancourt illuminates how the tattoos revealed during a sweat lodge ceremony—"faded ink / tapped in code"—reveal individual histories. In "Wedding Arch" (page 115), Lisa Stice reports a sharp surprise of welcome to the U.S. Marine Corps. In two separate works, each titled "Last Four," U.S. Army veteran Jessi M. Atherton (page 116) and U.S. Air Force veteran Lucas Randolph (page 159) label personal objects with snippets of Social Security Numbers. Finally, in "Gig Line" (page 8), U.S. Army and Coast Guard veteran Benjamin B. White comments on how trainees are taught to dress by vertically aligning the edges of a shirt with both the belt buckle and the trouser fly: "You would have had / to have been in the military / to know that."

Question No. 1
A "shibboleth" is something like a secret handshake, or a password. The word is defined as "a word or saying used by adherents of a party, sect, or belief." What is something—an item of clothing, a habit or mannerism, a symbol, a spoken phrase or response—that immediately reveals to you that someone has had experiences with the military?

Question No. 2:
How do shibboleths create opportunities for connection, empathy, and communication? Who do they include? Who do they exclude?

Writing Prompt:
Write about one or more habits, words, or beliefs—good or bad, useful or no longer useful—that you learned through your experiences with the military, or with some other organization. How do you find this item comforting, rewarding, or affirming? Alternatively, how do you find it disconcerting, harmful, or troubling?

Topic: "Things We Lost or Left Behind"

In "Wedding Ring" (page 108), U.S. Army veteran Chad Corrigan's poem marks the end of a third deployment with the toss of a wedding ring. In "Ode to My Skilcraft Pen" (page 74), U.S. Air Force veteran Eric Chandler eulogizes a favorite writing utensil. In Liam Corley's "Something Else You Don't Need" (page 76), the speaker's disposable pen seems to become a metaphor for something less-wanted. In "Hippie Car Stickers" (page 109), U.S. Navy spouse Andria Williams scrapes off bumper stickers to avoid an anticipated cultural clash.

Question No. 1
In what ways do we assess the "value" of an item?

Question No. 2:
How can losing and item sometimes be cathartic or therapeutic? How can intentionally (or even unintentionally) losing something help us control or move beyond difficult memories?

Writing Prompt:
Write about an item you wish you still had or one you are glad you have lost. Explore the memories connected with the item, and how those memories may have changed since the loss of the item. Explore the emotions (positive or negative) attached to that item and those memories.

Topic: "Bright Shiny Objects"

In Abby E. Murray's "Jewelry" (page 95), a bright, shiny gift becomes a symbol of a spouse's specific deployment—perhaps it is a peace offering? In Nancy Brown's "Jesus Hold My Earring" (page 97), a lost bauble may mirror a spouse's long absence from the household. In Mason Rodrigue's "Black Bracelets" (page 94), the engraved metal cuffs worn by some service members to commemorate fallen comrades seem to be at first desired, then dreaded. In contrast, in Rodrigue's "EGA" (page 46), the "Eagle, Globe & Anchor" emblem of the U.S. Marine Corps seems prized beyond measure.

Question No. 1:
How or why do we come to associate with jewelry—artful objects that arguably serve no other purpose than decoration—certain times, places, people, or achievements in our lives?

Question No. 2:
What types of jewelry, accessories, or decoration do you personally wear? Are any connected to your own military service, or the service history of someone you know or care about?

Writing Prompt:
Write about a piece of jewelry or decoration that you wear to evoke a certain time, place, person, or achievement. What do you say if and when people ask about the item? What do you keep secret, only for yourself?

Topic: "Boots, Boots, Boots"

In "Boots: *Origin < Old English*, remedy, fortunate" (page 9), U.S. Army veteran Karen Skolfield reflects on a gathering of beginner soldiers, and how they work to perfect a particular task. In "Boots on the Ground" (page 50), U.S. Marine veteran Aaron Graham illuminates what it means for a country to deploy its soldiers overseas. In "Boots in the Mud" (page 13), Vietnam War veteran Paul Hellweg describes a "soldier's cross"—a temporary, field-expedient memorial constructed of a rifle placed vertically into the ground, and topped by a helmet; at its base, a pair of empty boots.

Question No. 1:
What does it mean to "break-in" a pair of shoes, boots, or other footwear? What tricks or techniques have you used, or heard about using?

Question No. 2:
Do you have many shoes, or just a few prized pairs? What makes a pair of shoes special? Function? Looks? Quality of construction? Brand?

Writing Prompt:
Describe from memory a favorite pair of shoes, boots, or other footwear. Why were they favorites? How were they constructed? What functions did they offer? How did they make you feel, when you wore them? Where did they come from? Where did they go?

Topic: "Emotional Luggage"

In the 2012 movie "Memorial Day" (also titled "Souvenirs"), a 13-year-old boy on summer vacation discovers an old footlocker that is owned by his World War II veteran grandfather. His grandfather offers him a deal: Pick any three objects from the case, and he will tell the story behind each. In the poem "Footlocker" (page 55), U.S. Army veteran Colin D. Halloran describes the personal comforts tucked away in a footlocker being sent overseas to Afghanistan. In "Black Footlocker" (page 58), U.S. Marine veteran Aramis Calderon evokes the round-trip travels of a similar container. In Jehanne Dubrow's "Non-Essential Equipment" (page 35), the poet reveals items that will *not* be packed into a deploying sailor's sea bag. Finally, in Tony Mena's "So I Was a Coffin" (page 56), the poet imaginatively embodies a series of objects, becoming a vessel of various meanings.

Question No. 1:

What containers do you own or know about, that speak to military service, family heritage, or other personal history? Hope chests? Shadow boxes? Flag cases? What do they say about us—or the people who now possess them?

Question No. 2:

If you were to imagine yourself as an object or vessel, what shape would you take? From what materials would you be constructed? How would you be decorated?

Writing Prompt:

Write about a container of some sort, real or imagined. Start with sensory descriptions of the vessel, then consider: What objects does it contain? What memories does it evoke? Imagine removing three objects from the container: Describe each object with sense-provoking details. What does each object represent? What does the collection as a whole represent?

Topic: "Dulce et Decorum Est"

The classic war poem "Dulce et Decorum est" was written by Wilfred Owen during World War I, and published posthumously in 1920. The poem dispassionately describes the panic and horror of chemical warfare, with oft-quoted lines such as "Gas! GAS! Quick, boys!—An ecstasy of fumbling / Fitting the clumsy helmets just in time [...]" In his poem "CBR" (page 147)—the initialism stands for "Chemical-Biological-Radiological"—U.S. Army veteran Dale Ritter describes training with the M17A1 protective mask during the Vietnam War: "[K]nowing no matter / how good the training, we never learn anything, / that ignorance brings its own reward [...]" In his "Life and The First Gulf War, Sonnet No. 15" (page 151), U.S. Navy veteran Juan Manuel Pérez refers to the nuclear-biological-chemical training of the day as preparing for "the worst thing that could happen in a war," while also waiting in a foreign desert as legislators dicker back home.

Question No. 1: Protective masks and clothing are often recommended for such workaday tasks as sawing wood, applying spray paint, or working with household pesticides. Firefighters and medical personnel often use protective masks and equipment as part of their jobs.

In what ways do you use masks or similar equipment? (Gloves, aprons, face shields, safety glasses, etc.) What sensations, events, or memories do you associate with these uses?

Question No. 2: In what ways does protective equipment make us more safe? In what ways does protective equipment make us *less* safe?

Writing Prompt:
Describe an item or tool of protection, perhaps from memory or experience, or perhaps from your imagination. Describe the feelings you associate with that item.

Topic: "Rockets. Red. Glaring."

In the United States, fireworks are often associated with celebrations of freedom and independence, but the sights, sounds, and smells of explosives can also be ... *complicated* for veterans and their families. In "The Tranquil Fires in the Boat" (page 152), for example, U.S. Army veteran GOODW.Y.N. reacts to holiday fireworks that "explode like distant mortar rounds." She writes: "My mind is buried in the sands a million years past and away. / Where the celebrations were not so friendly, for they cherished war and nurtured death [...]" In contrast, in "Pen Flares" (page 153), U.S. Army veteran J.B. Stevens interrogates various memories he associates with signal flares that are shot from pen-sized launchers: "Laughter in a warzone is a precious thing (even if you might light your battle buddy on fire)." In "Ode to a Pineapple Grenade" (page 145), U.S. Marine veteran Gerardo "Tony" Mena writes, "[...] war is a performance, / and then we bow."

Question No. 1: Do you celebrate holidays such as Independence Day (U.S.) with fireworks, sparklers, or other explosive noise-makers? Why or why not?

Question No. 2: What memories, people, or events do you associate with loud and sudden noises? With "fireworks"?

Question No. 3: In what ways is a military performative?

Writing Prompt:
Describe an event that starts with a sizzle and ends with a bang. (Or a flash.) Or, alternatively, maybe one that doesn't go off as expected.

Topic: "Static Displays of Affection"

In "Snapshot" (page 92), poet Layle Keane Chambers describes a moment of anticipation and dread, captured in a phone's lockscreen image. In "The Great Valentine's Day Barrier Riff" (page 78), U.S. Army veteran David Abrams describes a dusty "I love you" spontaneously sent to his wife back home. In "Flags and Sunglasses" (page 61), U.S. Navy veteran Charles McCaffrey creates a subtle shrine to a fallen comrade—a picture on a desk, a pair of sunglasses nearby.

Question No. 1:
How and when do you take photographs? How and when do you share them with others?

Question No. 2:
How has your use or appreciation of photography evolved over time?

Writing Prompt:
Describe a scene, real or imagined, that is frozen frozen in photography. How was it made? Who is in the photograph? Who is missing? Where is the photograph located? How is it shared—or not shared—with others, and why?

Topic: "Fighting on One's Stomach"

An aphorism often attributed to Napoleon Bonaparte posits: "An army marches on its stomach." In "Pop Tarts, Rip-Its & the Surge" (page 130), however, helicopter pilot Chad Corrigan says the modern army marches on sugary treats and energy drinks. In Stan Gaidis's "Snack Times' 'Clicks and Snaps'" (page 156), the Vietnam veteran comes to associate certain sounds with a favorite wartime meal and refuge. In "Guacamole" (page 86), Iraq War veteran Jessi M. Atherton snarkily addresses a former love, through a recipe she has perfected.

Question No. 1:
"You are what you eat." How much of what we eat involves memory?

Question No. 2:
How many ways can we talk about food, without describing how it tastes?

Writing Prompt:
Write about a favorite food item or meal that connects to your military service, or the service of someone you know. Is it a recipe from Uncle Sam? Is it a taste of home? What memories does the food evoke? How would you describe the food to someone else, using each of the five senses? How would you describe the food, using each of the "five flavors"? (Sweet, salty, sour, bitter, and umami. If umami doesn't apply, try "spicy.")

About Lisa Stice

Lisa Stice is the author of the previously poetry collections *Uniform* (Aldrich Press, 2016), *Permanent Change of Station* (Middle West Press, 2018), *FORCES* (Middle West Press, 2021), and the poetry chapbook *Desert* (Prolific Press, 2018). Her work appears widely in literary journals and anthologies worldwide, the latter including *Beyond the Hill* (Lost Tower Publications, 2017); and *Nuclear Impact: Broken Atoms in Our Hands* (Shabda Press, 2017).

Stice's poem "Pursuit" was the 2020 military-family category poetry winner in the Col. Darron L. Wright Memorial Writing Awards, administered annually by the literary journal *Line of Advance*.

In 2017, her poem "Dear Wadih Sa'adeh" was selected as an honorable mention in the poetry category in that year's volume of the *Proud to Be: Writing by American Warriors* anthology series, published by Southeast Missouri State University Press. Her poem "A Quick Lunch from the Noodle Stand" was nominated for a 2016 Pushcart Prize by *The Magnolia Review*.

Stice has been an associate editor at Middle West Press LLC since 2023. She is also a poetry editor for *Inklette Magazine,* and often serves as an editor and mentor with various other writing organizations. The poet holds a Bachelor of Arts in English literature from Mesa State College (now Colorado Mesa University), Grand Junction, Colo., and a Master of Fine Arts in Creative Writing and Literary Arts from the University of Alaska, Anchorage.

She currently lives in North Carolina with her husband, daughter, and a beloved Norwich Terrier named Seamus.

You can learn more about her on-line at: lisastice.wordpress.com

Or follow her on Facebook and other social media: @LisaSticePoet

About Randy Brown

Randy Brown traveled the world as a child in an active-duty U.S. Air Force family in the 1970s, then landed permanently and happily in the American Midwest. A former editor of community and metro newspapers, as well as national trade and "how-to" consumer magazines, he is now a freelance writer and editor based in Central Iowa.

Brown embedded with his former Iowa Army National Guard unit as a civilian journalist in Afghanistan, May-June 2011. A 20-year military veteran with one overseas deployment, he subsequently authored the award-winning 2015 collection *Welcome to FOB Haiku: War Poems from Inside the Wire*. A chapbook, *So Frag & So Bold: Short Poems, Aphorisms & other Wartime Fun*, was published in 2021.

His poetry and essays have appeared widely in print and on-line, as well as anthologies. He even appeared as an "on screen" character in the 2021 *True War Stories* anthology from Z2 Comics, Denver.

Brown is a three-time poetry finalist in the Col. Darron L. Wright Memorial Writing Awards. He co-edited the 2019 Military Writers Guild anthology *Why We Write: Craft Essays on Writing War*, and curated the 2015 *Reporting for Duty: U.S. Citizen-Soldier Journalism from the Afghan Surge, 2010-2011*.

Brown was the winner of the 2018 "Untold Stories" poetry contest administered by *Flyover: Journal of Writing & the Environment*. He was the 2015 winner of the inaugural Madigan Award for humorous military-themed writing, presented by Negative Capability Press, Mobile, Alabama.

He is the current poetry editor at the literary journal *As You Were*, published twice a year by the non-profit Military Experience & the Arts. He is also a member of Military Reporters & Editors, the Military Writers Guild, and the Military Writers Society of America.

As "Charlie Sherpa," he writes about modern war poetry at: www.fobhaiku.com; and military writing at: www.aimingcircle.org.

About Vicki Hudson

Victoria A. Hudson creates poetry, photography, narrative non-fiction, and occasional fiction. She also coaches rugby.

Hudson retired from the United States Army in December 2012, serving 33 years, one month and 15 days. During her career, she held command, staff, and line assignments in military police, civil affairs, and information operations. She is a proud prior-enlisted "Mustang" and loves to wear her 11th Cavalry Stetson while out and about. In Iraq Rotation 3 (2005), she was a convoy commander with a Civil Affairs battalion operating in the Mosul Area of Responsibility (A.O.R.).

Her writing has appeared in such venues as *Outserve Magazine, O Dark Thirty, Milvia Street,* and *Addana Literary Journal.* An essay and poem were used as part of the source material for the stage play *Coming in Hot,* an adaptation of the 2008 book *Powder: Writing by Women in the Ranks, from Vietnam to Iraq,* published by Kore Press.

In 2011, her essay "Why Our Family is Fighting DOMA [Defense of Marriage Act]" appeared in the November issue of the *Advocate.* Her story was part of an amicus brief submitted to the U.S. Supreme Court in United States v. Windsor.

She is the author of the 2012 non-fiction book, *No Red Pen— Writers, Writing Groups & Critique,* a manual for individuals, writing groups, and teachers for insights into developing inclusive, respectful workshopping processes.

Hudson holds a Master of Fine Arts in nonfiction from Saint Mary's College of California. She also holds a graduate degree in sport and performance psychology from John F. Kennedy University, Pleasant Hill, California.

She lives in Northern California with her wife and two children, an army of cats, and one failed service dog. You can find her often in Azeroth as McVicster#1973.

Learn more about her at: www.vickihudson.com

www.ingramcontent.com/pod-product-compliance
Lightning Source LLC
Chambersburg PA
CBHW030526020726
47494CB00004B/1247